Elements of Rock Physics and Their Application to Inversion and AVO Studies

T0144444

Elements of Rock Physics and Their Application to Inversion and AVO Studies

Robert S. Gullco and
Malcolm Anderson

CRC Press
Taylor & Francis Group
Boca Raton London New York

CRC Press is an imprint of the
Taylor & Francis Group, an **informa** business

First published 2022
by CRC Press/Balkema
Schipholweg 107C, 2316 XC Leiden, The Netherlands
e-mail: enquiries@taylorandfrancis.com
www.routledge.com - www.taylorandfrancis.com

CRC Press/Balkema is an imprint of the Taylor & Francis Group, an informa business

Library of Congress Cataloging-in-Publication Data
Names: Gullco, Robert S., author. | Anderson, Malcolm (Mathematician), author.
Title: Elements of rock physics and their application to inversion and AVO studies /
Robert S. Gullco and Malcolm Anderson.
Description: Leiden, The Netherlands ; Boca Raton :
CRC Press/Balkema, 2022. | Includes bibliographical references.
Subjects: LCSH: Amplitude variation with offset analysis. | Rock deformation–Mathematical models. | Seismology–Mathematics. | Seismic reflection method. | Seismic prospecting. | Geology, Structural.
Classification: LCC QE539.2.S43 G85 2022 (print) | LCC QE539.2.S43 (ebook) |
DDC 620.1/125–dc23/eng/20211029
LC record available at https://lccn.loc.gov/2021041377
LC ebook record available at https://lccn.loc.gov/2021041378

ISBN: 978-1-032-19993-1 (hbk)
ISBN: 978-1-032-13495-6 (pbk)
ISBN: 978-1-003-26177-3 (ebk)

DOI: 10.1201/9781003261773

Typeset in Times New Roman
by codeMantra

Contents

About the authors

Robert S. Gullco received his Master's degree in Geological Sciences from the University of Buenos Aires (Argentina). He has worked as a geologist and petrophysicist in YPF (then Argentina state oil company), Wapet (now Chevron) in Western Australia, Paradigm Geophysical (in both Australia and Mexico), and CGG and Citla Energy in Mexico. He has taken courses on geomathematics at Stanford University and a course on applied mathematics at Curtin University.

Malcolm Anderson took his Master's degree in Applied Mathematics and Theoretical Physics at the Australian National University in Canberra before completing a PhD in Theoretical Astrophysics at the Institute of Astronomy in Cambridge. He has taught applied mathematics and mathematical physics at the Australian National University, the University of New South Wales and Edith Cowan University. Since 2000, he has been a member of the Mathematics Group at Universiti Brunei Darussalam.

Introduction

In principle, the terms "rock physics" and "petrophysics" are synonymous. In practice, however, petrophysics refers more to the theory and techniques used in formation evaluation, which include well logs, cores and pressure tests.

The aim of a petrophysics analysis is to determine the porosity, water saturation and permeability in a sequence at well scale, in order to choose the best intervals for perforation, to estimate the original volume of hydrocarbons in place in the reservoir, to support any reservoir simulation exercise and to optimize production. Petrophysics, in this sense, requires the collaboration of a reservoir engineer and a "petrophysicist" (who is generally a geologist, a geophysicist or an engineer).

Rock physics, on the other hand, is generally understood to be the study of those physical properties of rocks which are relevant to seismic exploration. If an exploration team has a certain idea of the physical properties of the rocks in a given area, it is possible for them to interpret the seismic data in the area (which are typically the results of an inversion or an AVO, Amplitude Variation with Offset) in terms of lithologies and fluid content. This type of rock physics analysis requires the collaboration of a geophysicist and a petrophysicist.

The ultimate aim of rock physics is very ambitious: to determine, for a certain body of rock often buried deep underground, the lithology, the porosity and the fluid content at every point in the body from indirect measurements of the P-velocity, the S-velocity and the density. In practice, the results of seismic prospecting are usually much more modest. However, these results can be improved if in the area under consideration there is at least one well with a complete set of logs.

The scope of rock physics can be as broad as the expertise of the "rock physicist" (who is generally a petrophysicist or a geologist, etc.) allows. In this book, we present a detailed explanation – in many cases with real-life examples – of the topics included in the description of the discipline mentioned above. Other topics that are important to petroleum engineering, such as wellbore stability, will not be considered here.

The mathematics needed to describe a non-spherical seismic wave travelling through an anisotropic and heterogeneous medium is extremely complex. Only if we make simplifying assumptions and adopt very simple models we can develop practical methods of analysis for the rock physicist. The use of simplified models and empirical equations avoids many of the mathematical complications, but the price to be paid is a degree of uncertainty in all the inferences that are made.

DOI: 10.1201/9781003261773-1

This book is intended primarily for practitioners of rock physics who are in possession of seismic data (the results of inversion and/or AVO) and want to translate these data into information about porosity, lithology and fluid content. The book should also be of interest to students in all disciplines of the geosciences.

Chapter 1

Petrophysics review

DEFINITION OF EFFECTIVE AND TOTAL POROSITY, CLAY AND SHALE

When dealing with clastic stratigraphic columns, we will generally make use of rock models such as the effective porosity/shale model or the total porosity/dry clay model. It is therefore very important to clarify what is meant in this book by "effective porosity", "total porosity", "shale" and "clay". Some of the definitions given below are not universally accepted. When a rock physics study is performed to interpret inversion or AVO (Amplitude Variations with Offset) data, the well log evaluation (if any well is present) has already been done. In such a situation, it is important for the analyst to check with the petrophysicist who evaluated the well the meanings of the various diagnostic categories being used, such as "porosity" (which will sometimes appear in a LAS file – special type of ASCII file most suitable to store well log data – as a column labelled just PHI) or "Vclay".

In clean rocks (*i.e.* rocks devoid of clay), there is no ambiguity regarding the meaning of the term "porosity", which we will here denote by the symbol ϕ. It is simply the pore volume V_p divided by the bulk volume V_b of the rock. In symbols:

$$\phi = \frac{V_p}{V_b} \tag{1.1}$$

For clean rocks, the effective porosity and the total porosity are identical and can be calculated from equation (1.1). However, even in clean rocks, there may be pores that are isolated and do not contribute to the movement of fluids through the rock. For a reservoir engineer, equation (1.1) measures the "total porosity", whereas the value of the "effective porosity", which includes only those pores that are interconnected, will of course be lower than the "total porosity". It could be argued that there is a certain relationship between the irreducible water saturation, which can be calculated from capillary pressure curves, and the non-connected porosity. However, these are somewhat different concepts. The irreducible water saturation refers to that part of the pore space composed of capillaries so fine that very high pressures would be needed to displace the water with another fluid. So, by definition, it is a property of the connected pores.

In this book, we will follow the standard petrophysics convention that, in clean rocks whose porosity has been calculated with well logs, the effective and total porosities are the same.

When a clastic rock is not "clean", there are two principle alternative models that can be used to describe it.

DOI: 10.1201/9781003261773-2

The effective porosity model

To understand this model, it is necessary to first define what is meant by "shale". Geologically, shale is defined as a very fine-grained rock, whose particles are predominantly of clay size (less than 1/256 mm), but may include some of silt size (less than 1/16 mm). A "geological" shale is finely laminated and is composed of clay minerals (illite, kaolinite, montmorillonite, etc.), quartz and feldspars. Table 1.1 lists the approximate mineralogical composition of "shale" according to different researchers:

In a petrophysics context, the definition of shale is somewhat different from the geological one. Although "shale" still refers to a collection of very fine particles with a high proportion of clay minerals and quartz, it is considered to be a single (hypothetical) mineral, with the properties of the mixture of its constituents. This mineral has its own porosity, which is a measure of the pore volume between the fine particles constituting the "shale" and by water adsorbed by the clay minerals. It is conventional to regard the pore space in shale as unconnected and so as absolutely impermeable. The terms "shale" and "clay" are definitely not synonymous in either petrophysics or geology.

The effective porosity model includes two solid components: the "matrix" (which generally has the physical properties of quartz) and the "shale" (whose physical properties are calculated from well logs). In addition, the voids (the "effective" or permeable pore space) may contain hydrocarbons and water. The saturation of the hydrocarbons is defined to be V_h / V_p (*i.e.* the volume of hydrocarbons divided by the pore volume), while the effective water saturation is $S_{we} = V_w / V_p$ (the volume of water in the voids divided by the pore volume).

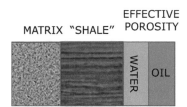

Figure 1.1 Schematic view of the "effective porosity model".

Table 1.1 Estimated mineralogical composition (per cent) of shale according to several authors

Components	Leith & Mead	Clarke	Krynine	Yaalon
	1915	1922	1948	1961
Clay Minerals	34	25	47	59
Quartz	32	22.3	29	20
Feldspars	18	30	5	8
Carbonates	8	5.7	7	7
Iron oxides	5	5.6	6	3
Others	2	11.4	6	3

In the effective porosity model,

$$V_{qtz} + V_{sh} + \phi_e = 1 \tag{1.2}$$

where V_{qtz} is the volume fraction of quartz, V_{sh} is the volume fraction of shale and ϕ_e is the effective porosity (*cf.* Figure 1.1).

Despite its simplicity, this model accurately represents well log evaluations, and it seems to also work well in rock physics problems.

The total porosity model

This model has several variants, but we will consider just one of them here. Assume for the sake of simplicity that the rock contains only two minerals, one being quartz and the other a specific clay species (say illite) whose physical properties are relatively well known. The clay mineral is treated as a "dry" component, and the water bound to it is included in the "total porosity". The amount of water bound to the clay fraction can be calculated as a function of the temperature and water salinity. This bound water, together with any free water and hydrocarbon content, constitutes the total porosity of the rock. In the total porosity model, therefore

$$V_{qtz} + V_{clay} + \phi_t = 1 \tag{1.3}$$

where V_{qtz} is the volume fraction of quartz as before and V_{cldry} is the volume fraction of dry clay (*cf.* Figure 1.2).

The total porosity ϕ_t in equation (1.3) is given by:

$$\phi_t = V_{bw} + V_{fw} + V_{hyd} \tag{1.4}$$

where V_{bw} is the volume fraction of bound water, V_{fw} is the volume fraction of free water and V_{hyd} is the volume fraction of hydrocarbons. It follows that the total water saturation is:

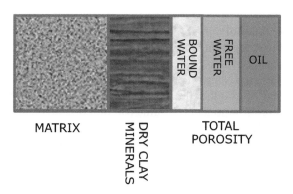

Figure 1.2 Schematic view of the total porosity model.

$$S_{wt} = \frac{V_{bw} + V_{fw}}{V_{bw} + V_{fw} + V_{hyd}} = \frac{V_{bw} + V_{fw}}{\phi_t} \qquad (1.5)$$

An important relationship linking the effective porosity and water saturation with the total porosity and water saturation is:

$$\phi_t \left(1 - S_{wt}\right) = \phi_e \left(1 - S_{we}\right) \qquad (1.6)$$

It is clear that the effective porosity model is simpler than the total porosity model. Nonetheless, the latter has a great appeal in the petrophysics community, because the standard methods for calculating water saturation in "dirty" sands, using either the dual water formulae or the Waxman and Smits formulae, have a solid theoretical background. In the effective porosity model, the water saturation in "dirty" sands is calculated by means of empirical formulae, such as the Indonesia or Simandoux formulae. Despite the differences, the final results are quite similar and it is difficult to determine which of the methods produces the best results.

In the experience of the authors, it is simpler in rock physics applications to work with the effective porosity (or shale) model. As will be seen in later sections, it is very simple to estimate the elastic parameters of the shale from well log data. Hence, if a well log evaluation has already been carried out using the total porosity method, it may be more practical to convert the total porosity into an effective porosity and the total water saturation into an effective water saturation. An LAS file containing the results of an evaluation carried out using the total porosity model will usually include columns of values of "Vbw" (the volume fraction of bound water), "Vdcl" (the volume fraction of dry clay) or "Vwcl" (the volume fraction of wet clay), in addition to the columns labelled "Swt" and "PHIT". Although at a conceptual level they are not strictly the same, the volume fraction of wet clay can be identified with the shale volume fraction, and the effective porosity is then equal to the total porosity minus the volume fraction of bound water. The effective water saturation in this case can be calculated from equation (1.6).

ESTIMATION OF THE SHALE POINT IN A DENSITY/NEUTRON CROSSPLOT

Because of the importance of the "shale" concept, we will illustrate how the shale point is determined from well log data. This operation is a fundamental step in the well log evaluation of a clastic rock.

Figure 1.3 is a Density/Neutron crossplot, which in this particular case comprises data from an interval of about 1200 m. In the effective porosity model governed by equation (1.2), the three components (volume fraction of quartz, effective porosity and volume fraction of shale) add up to one. Two points are already fixed in this crossplot: the quartz point (density = 2.65 g/cc, and quartz hydrogen index about –0.05 v/v for a limestone-calibrated Neutron log) and the water point (density = 1.0 g/cc and water hydrogen index = 1.0 v/v). The quartz and water points are marked by circles in the diagram.

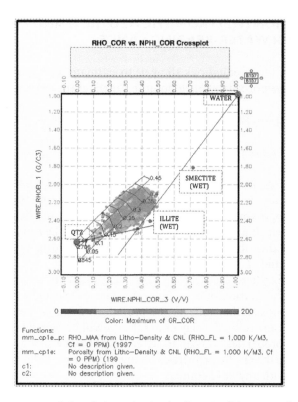

Figure I.3 Determination of the shale point in the Density/Neutron plane.

What is not known beforehand is the shale point, which needs to be determined. According to the model, all the observed points should be enclosed by a triangle, so that equation (1.2) is satisfied at each point, and the volume fractions of the three components are all between 0 and 1. To find the shale point, we first trace a line which passes through the water point and rotate it so that a great majority of the observed points are above the line. In other words, the line should be approximately tangential to the observed points. We then trace a second line passing through the quartz point and rotate it so that a great majority of the observed points are also above the line. The intersection of the two lines determines the shale point (shown in Figure 1.3 as the point marked "SH"). Once the location of the shale point is known, we can calculate the volume fraction V_{sh} and the effective porosity for all the points in the diagram. If gas was present, any sand points with low V_{sh} would fall above the quartz/water line.

Note that in Figure 1.3 the positions of the "wet smectite" and "wet illite" points are also indicated. In principle, the volume of water bound to a specific clay type can be calculated as a function of the temperature and the water salinity. Once the volume fraction of the water bound to, say, the illite fraction is known, we can calculate the Density and Neutron responses of the wet clay. In Figure 1.3, we can see that the illite point inferred from the temperature and salinity data is quite close to the shale point estimated from the diagram.

CALCULATION OF THE EFFECTIVE POROSITY AND THE SHALE VOLUME FRACTION (V_{SH}) FROM THE NEUTRON AND DENSITY LOGS, IN OIL- OR WATER-BEARING SANDS

The following set of linear equations allow us to estimate the volume fraction of shale (V_{sh}) and the effective porosity at each point in the diagram:

$$\rho_b = \rho_{ma}\left(1 - V_{sh} - \phi_e\right) + \rho_{sh}V_{sh} + \rho_f\phi_e \tag{1.7}$$

$$\phi_N = H_{ma}\left(1 - V_{sh} - \phi_e\right) + H_{sh}V_{sh} + H_f\phi_e \tag{1.8}$$

In these equations, ρ_b and ϕ_N are the Density and Neutron tool responses, respectively, ρ_{ma} and H_{ma} are the density and hydrogen index of the matrix (for quartz, these are 2.65 g/cc and approximately –0.05 v/v), ρ_{sh} and H_{sh} are the density and hydrogen index of the shale (which are fixed by the shale point in the Density/Neutron crossplot in Figure 1.3), while the subscript "*f*" denotes the corresponding properties of the fluid (for pure water, ρ_f and H_f are 1 g/cc and 1 v/v, respectively).

It bears repeating here that "shale" is strictly speaking a hypothetical mineral, which in practice is a mixture of clay minerals, quartz, feldspars and other minerals, and has its own porosity that reflects the presence of spaces between the very fine grains and water adsorbed by the clay minerals. As will be seen later, this hypothetical mineral has elastic properties that can be determined from well logs. Furthermore, it is possible to calculate the elastic properties of any mixture of shale, quartz and effective porosity, together with any fluid content.

USING THE GAMMA RAY LOG TO CALCULATE SHALE VOLUME FRACTION: COMPARISON WITH THE NEUTRON/DENSITY APPROACH

It is generally accepted that clean sands have a low level of radioactivity, whereas shales are much more radioactive, due to the presence of clay minerals and feldspars, which contain potassium and sometimes thorium and uranium as impurities. If a histogram is constructed of the Gamma Ray (GR) measurements over the interval of interest, the two extreme values are the most critical: The lowest response defines GR_{ma} (the GR response of the matrix) and the highest response defines GR_{sh} (the GR response of the shale). If we assume a linear relationship between GR and V_{sh}, we can calculate V_{sh} directly from GR by using the simple formula:

$$V_{sh} = \frac{GR - GR_{ma}}{GR_{sh} - GR_{ma}} \tag{1.9}$$

However, there are some problems in using the GR log to estimate V_{sh}. First, the two extreme values, GR_{ma} and GR_{sh}, are somewhat arbitrary. Assigning $V_{sh} = 0$ to the minimum value of the GR log over the interval of interest is equivalent to assuming that just one point in the column is clean sand, completely devoid of shale. On the other hand, if GR_{ma} is instead taken to be the cumulative value of GR at the bottom 5% of points in the histogram, the default assumption is that 5% of the points are absolutely clean sands. A problem that is potentially even more serious is that the matrix might not be pure quartz, as it may contain feldspar as well, in which case sands devoid of shale could still have a high GR value. For these reasons, it is preferable to calculate V_{sh} using the Neutron and Density logs, following equations (1.7) and (1.8). However, in gas-bearing sands there are too many unknowns, and the Gamma Ray log is also needed to estimate V_{sh}. Gas-bearing sands can be identified visually in well logs (see the next section). In intervals of this type, V_{sh} should be estimated from the GR log, using equation (1.9). The reliability of the results obtained from the GR log in this way can be checked as follows.

If the effective porosity ϕ_e is eliminated from equations (1.7) and (1.8), these reduce to the single relation

$$Vsh = \frac{\dfrac{\rho_b - \rho_{ma}}{\rho_f - \rho_{ma}} - \dfrac{\phi_N - H_{ma}}{H_f - H_{ma}}}{\dfrac{\rho_{sh} - \rho_{ma}}{\rho_f - \rho_{ma}} - \dfrac{H_{sh} - H_{ma}}{H_f - H_{ma}}} \tag{1.10}$$

Note that the numerator in equation (1.10) is a known quantity if the Density and Neutron logs are available. If X denotes the numerator and k is the reciprocal of the denominator, equation (1.10) reads:

$$V_{sh} = kX \tag{1.11}$$

After replacing V_{sh} in (1.9) with kX and solving for GR, we get:

$$GR = GR_{ma} + k(GR_{sh} - GR_{ma})X \tag{1.12}$$

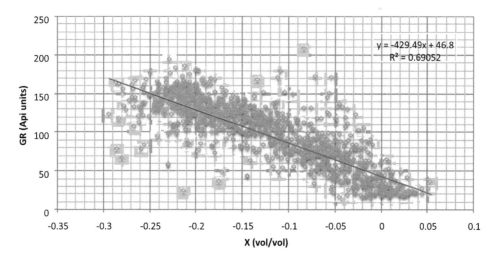

Figure 1.4 A plot of GR vs. X. Data are taken from a sand/shale sequence, about 175 m thick. Note the high correlation coefficient between the variables, which suggests that the model described by equations (1.7)–(1.9) is reasonable. Because the regression line corresponds to $GR = GR_{ma} + k(GR_{sh} - GR_{ma})X$, both GR_{ma} and GR_{sh} can be calculated from the coefficients of the line.

In theory, if we were to plot the values of GR against X, the points should lie on a straight line, with GR_{ma} the intercept on the GR axis. However, the points will in practice never fall on a perfect straight line. Figure 1.4 shows a plot of GR vs. X for a 175 m interval in a sand/shale sequence. Note the relatively high correlation coefficient. The fact that the points do approximate a straight line reasonably well suggests that equations (1.7), (1.8) and (1.9) are modelling the lithology over this interval with reasonable accuracy. On the other hand, a low correlation coefficient could imply, for instance, that the sands are radioactive and so the GR log is not suitable for estimating the shale volume fraction.

If we do obtain a high correlation between GR and X, we can be confident that the values of V_{sh} calculated using the Gamma Ray log are accurate for gas-bearing intervals (and furthermore, the intervals containing gas can be identified visually). The parameters GR_{ma} and GR_{sh} can then be taken from the GR vs. X crossplot, because the value of k is already known. However, if there is a poor correlation between GR and X, the GR log should not be used to estimate V_{sh} even for gas-bearing intervals.

EVALUATING GAS-BEARING SANDS USING THE NEUTRON, DENSITY AND GAMMA RAY LOGS

As a first step, we will comment on the visual identification of gas-bearing sands in a well.

Figure 1.5 shows log data from a sand/shale sequence about 100 m thick. The first track on the left is the Gamma Ray log, followed by two depth tracks. The induction log follows, and to the right of that is a track containing the Neutron and Density logs. The Neutron log has a scale running from 0.60 (on the left) to 0 (on the right), while the scale for the Density log varies from 1.65 g/cc (on the left) to 2.65 g/cc (on the right). Finally, the track on the extreme right contains the two sonic logs.

It is evident from Figure 1.5 that the section between 3202 m and 3231 m has high resistivity and relatively low GR values. These are characteristics of hydrocarbon-bearing sand. In the track containing the Density and Neutron logs, the Neutron response lies to the left of the Density value over most of the interval shown. However, throughout the 3202–3231 m section, the positions of the two curves are reversed, and the space between the curves in this "crossover" region has been shaded. The crossover indicates that the interval contains gas. The Neutron response in gas-bearing formations is very low, and the density is also low relative to the column average, hence the crossover. Of course, the scales of both logs also play a role in the crossover. The scale shown in this well for the two logs is fairly common. Another very common scale would be from 0.45 v/v (on the left) to −0.15 v/v (on the right) for the Neutron log, and from 1.95 g/cc (on the left) to 2.95 g/cc (on the right) for the Density log. Either of these scales would typically show a crossover if gas is present. Whenever a gas zone is identified, it is not possible to make a quantitative evaluation using the Neutron/Density logs alone. The Gamma Ray log is used exclusively to calculate V_{sh} (prior to making any other calculation), provided that there is a high correlation coefficient in the plot of GR vs X. The Gamma Ray log is needed to calculate V_{sh} because the density and the hydrogen index of the fluid cannot be assumed to have their default values, which are 1 g/cc and 1 v/v, respectively, when gas is present. Instead, they are unknowns whose values need to be estimated separately.

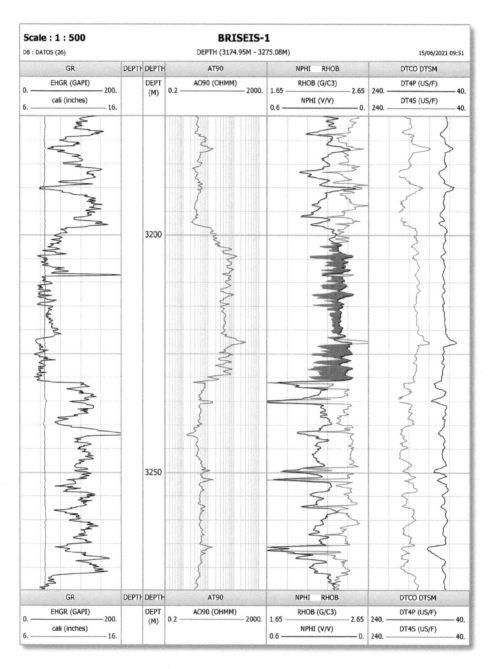

Figure 1.5 Identification of a gas zone over the interval 3202–3231 m, indicated by the crossover of the Neutron and Density logs (shaded).

In the case where gas is present in the formation, the following equations need to be solved with the help of the Density/Neutron data:

$$\rho_b = \rho_{ma}(1-\phi_e - Vsh) + \rho_{sh}Vsh + \rho_f\phi_e \tag{1.13}$$

$$\phi_N = H_{ma}(1-\phi_e - Vsh) + H_{sh}Vsh + H_f\phi_e - \left[(1-H_f)(2H_f\phi_e^2)\right] \tag{1.14}$$

In equation (1.14), the term in square brackets represents the so-called "excavation effect" (see Serra, 1984), which reduces the magnitude of the Neutron reading in gas-bearing rocks. Note that in water sands, and in oil sands as well, the value of H_f (the fluid hydrogen index) is close to 1 and the "excavation effect" term can be ignored, as was done for example in equation (1.8).

At this stage, it is understood that the value of V_{sh} has already been calculated using the Gamma Ray log and therefore is no longer unknown. The density ρ_f and hydrogen index H_f of the fluid can be calculated from the relations:

$$\rho_f = \rho_w S_{xo} + (1 - S_{xo})\rho_{gap} \tag{1.15}$$

$$H_f = H_w S_{xo} + (1 - S_{xo})H_g \tag{1.16}$$

where S_{xo} is the saturation of the mud filtrate in the invaded zone (*i.e.* the vicinity of the borehole, the region from which the Neutron and Density data are taken) and $1 - S_{xo}$ represents the gas saturation in this zone. Also, ρ_{gap} in equation (1.15) is the apparent gas density, as read by the Density tool. The apparent gas density is related to the true gas density by the equation

$$\rho_{gap} = 1.335\rho_g - 0.188 \tag{1.17}$$

The parameter S_{xo} is an unknown quantity in a gas-bearing interval, which is why the GR log is needed to calculate V_{sh}. Otherwise, there would be just two equations (involving the responses of the Neutron and the Density logs) for the three unknowns V_{sh}, the effective porosity and S_{xo}. The actual values of the gas density and hydrogen index will depend on the characteristics of the reservoir being explored. In Appendix 1.1, it is explained how the hydrogen index and density for a gas (methane) can be calculated at any pressure and temperature.

If equation (1.15) is used to eliminate S_{xo} from equation (1.16), the equation that results is

$$H_f = \left[H_g - \left(\frac{H_w - H_g}{\rho_w - \rho_{gapp}}\right)\rho_{gap}\right] + \left[\frac{H_w - H_g}{\rho_w - \rho_{gapp}}\right]\rho_f \tag{1.18}$$

In equation (1.18), the two quantities enclosed in the square brackets are constants, so the equation can be represented schematically in the form

$$H_f = A + B\rho_f \tag{1.19}$$

Equations (1.13) and (1.14) can also be rewritten as

$$(\rho_b - \rho_{ma}) - Vsh(\rho_{sh} - \rho_{ma}) = \phi_e(\rho_f - \rho_{ma}) \tag{1.20}$$

$$\left(\phi_N - H_{ma}\right) - Vsh\left(H_{sh} - H_{ma}\right) = \phi_e\left(H_f - H_{ma}\right) - \left(1 - H_f\right)\left(2\phi_e^2 H_f + 0.04\phi_e\right) \quad (1.21)$$

In both equations (1.20) and (1.21), the expressions on the left-hands of the equations are known quantities. So, the equations can be represented schematically as

$$P = \phi_e\left(\rho_f - \rho_{ma}\right) \qquad (1.22)$$

$$Q = \phi_e\left(H_f - H_{ma}\right) - \left(1 - H_f\right)\left(2\phi_e^2 H_f + 0.04\phi_e\right) \qquad (1.23)$$

If equation (1.22) is used to eliminate ϕ_e and equation (1.19) to eliminate H_f, from equation (1.23), the resulting equation reads

$$Q = \frac{P(A + B\rho_f - H_{ma})}{(\rho_f - \rho_{ma})} - (1 - A - B\rho_f)\left[\frac{2(A + B\rho_f)P^2}{(\rho_f - \rho_{ma})^2} + 0.04\frac{P}{(\rho_f - \rho_{ma})}\right] \quad (1.24)$$

Equation (1.24) is effectively a quadratic equation $a\rho_f^2 + b\rho_f + c = 0$ for ρ_f, which solves to give:

$$\rho_f = \frac{-b \pm \sqrt{b^2 - 4ac}}{2a} \qquad (1.25)$$

where the coefficients a, b and c can be calculated directly from equation (1.24). That is, after multiplying both left and right sides of equation (1.24) by $(\rho_f - \rho_{ma})^2$ and collecting all the terms together on one side of the equation, a is the algebraic sum of the coefficients of the terms containing the square of the fluid density ρ_f, b is the sum of coefficients of the terms linear in ρ_f and c is the sum of the terms independent of ρ_f.

Once the value of the fluid density ρ_f has been calculated, equation (1.15) can be used to solve for S_{xo} (the water or mud filtrate saturation in the invaded zone) and equation (1.22) to solve for the porosity ϕ_e. Although S_{xo} is not the same as S_w (the water saturation in the virgin zone of the well), the value of S_{xo} estimated in this way is independent of any electric measurements.

REFERENCE

Serra, O. (1984), *Fundamentals of Well Log Interpretation*, Elsevier, Amsterdam-Oxford-New York-Tokyo, Third Impression 1988.

APPENDIX 1.1: GAS DENSITY AND HYDROGEN INDEX AT RESERVOIR CONDITIONS

Hydrogen index of a gas

The hydrogen index of any substance is defined as

$$HI = \frac{No.\ of\ atoms\ of\ Hydrogen\ in\ 1cm^3\ of\ substance\ at\ reservoir\ conditions}{No.\ of\ atoms\ of\ Hydrogen\ in\ 1\ cm^3\ of\ pure\ water\ at\ standard\ conditions}$$

The value of the denominator in this equation is 0.669×10^{23} atoms.

Let us assume that a mixture of gases contains a total of n moles of molecules in 1 cc at reservoir conditions. The total number of molecules in n moles of gas is $6.02 \times 10^{23} n$, so the total number of hydrogen atoms in 1 cc of the mixture is given by:

$$6.02 \times 10^{23}(4n_1 + 6n_2 + 8n_3 + \ldots.),$$

where n_1 is number of moles of methane and n_2 is the number of moles of ethane, etc.

If $x_k = n_k / n$ is the molar fraction of the kth hydrocarbon, after dividing this expression through by 0.669×10^{23}, the formula for the hydrogen index becomes

$$HI = 8.9985n(4x_1 + 6x_2 + 8x_3 + \ldots...) \tag{1}$$

But the total number of moles, n, is just the mass of the gas divided by the molecular weight of the gas, where the mass of the gas can be represented symbolically as $\rho_g V_g$. Furthermore, the gas volume V_g in this case is $1\,\mathrm{cm}^3$. So, equation (1) becomes:

$$HI = 8.9985\rho_g \frac{(4x_1 + 6x_2 + 8x_3 + \ldots...)}{(M_1 x_1 + M_2 x_2 + M_3 x_3 .....)} \tag{2}$$

where M_k is the molecular weight of the kth hydrocarbon in the mixture.

Note here that the hydrogen index is proportional to the gas density, which in turn is strongly dependent on the pressure and temperature of the reservoir.

In the particular case of pure methane, whose molecular weight is 16 g, equation (1) with $x_1 = 1$ reduces to:

$$HI = 2.2496 \, \rho_g \tag{3}$$

Gas density at reservoir conditions

This topic will be revisited in Chapter 5, where the problem of fluid substitution is examined.

The dependence of the gas density on the reservoir conditions can be modelled by the approximate ideal gas law

$$\rho_g = \frac{pM_w}{ZRT} \tag{4}$$

where p is the pressure (in atmospheres), M_w is the molecular weight (in grams), T is the temperature (in K), and R is the gas constant [0.082 L atm/(K Mole)]. Z is a factor that accounts for deviations from the behaviour of an ideal gas. If as a first approximation the gas is assumed to be ideal, $Z = 1$, for all temperatures and pressures. However, it is always advisable to take the deviation factor into account. There are several methods available for estimating the density of a real gas. As an example, the following table shows a set of calculations taken from Mavko et al. (1998, pp. 216–217).

Table 1.1.1 shows values of the density of pure methane for a set of pressures and temperatures likely to be encountered in real reservoirs. The calculations were

Table 1.1.1 Density of pure methane for a set of pressures and temperatures likely to be encountered in real life.

Temp (°C)	Pr. (kgf/cm²)	ρg (g/cc)	ρgap (g/cc)	HI (admi.)
18	53	0.0385	−0.1366	0.0866
24	79	0.0581	−0.1104	0.1307
30	105	0.0769	−0.0853	0.1730
36	144	0.1043	−0.0488	0.2346
43	173	0.1208	−0.0267	0.2718
49	201	0.1352	−0.0075	0.3041
55	230	0.1483	0.0100	0.3336
61	259	0.1597	0.0252	0.3593
68	300	0.1743	0.0447	0.3921
74	330	0.1828	0.0560	0.4112
80	360	0.1902	0.0659	0.4279
86	390	0.1967	0.0746	0.4425
93	420	0.2017	0.0813	0.4537

The apparent gas density (as read by the Density tool) and the hydrogen index are also shown in the table.

performed using the method explained in this section. The apparent gas density (as read by the Density tool) and the hydrogen index were also calculated, and appear in the fourth and fifth columns, respectively.

Chapter 2

Elements of elasticity theory

DEFINITION OF STRESS, STRAIN, ELASTICITY AND ELASTIC MODULI

When energy is injected into a material body, the energy propagates in all directions, but there is no net displacement of the matter itself. The particles that make up the material body (which would otherwise be motionless) oscillate slightly about their equilibrium positions. If the injection of energy is only transient, the oscillatory movement of the particles will eventually be damped away. This flux of energy, in conjunction with the oscillation of the particles, is conventionally called a "wave".

For our purposes, there are two types of body waves (in addition to surface waves). The primary waves are compressional or longitudinal waves. In this type of wave, the movement of the particles is parallel to the direction of propagation of the wave. The second type of body wave is the shear or transverse wave, in which the oscillations of the particles are perpendicular to the direction of propagation of the wave (similarly to an electromagnetic wave).

In the derivations that follow, we will assume that the rocks are homogenous, isotropic and elastic. The first task therefore is to define what is meant by isotropy and elasticity.

A material body is said to be "isotropic" if all its physical properties are independent of direction. In a stratified lithological column, the velocities along the bedding plane can be quite different to the velocities normal to the bedding plane, which means that the column is not strictly isotropic. This is a problem rock physicists simply have to live with. If we were to take into account the anisotropy of the rock, the resulting elasticity equations would be extremely complex.

If a stress (a force per unit area) is applied to a material body, the body will deform to a certain degree. If the body returns to its original shape when the stress is withdrawn, we say that the body is "elastic".

When a wave passes through a body and the individual particles begin to oscillate, a deformation of the body is involved. During the transmission of seismic waves, the deformations involved are so insignificant that the body quickly returns to its original shape. As a happy consequence of this, rocks tend to behave as elastic solids when disturbed by seismic waves.

If an elastic band with natural length L is stretched to the point where its length increases by an amount Δl, the quantity $\Delta l/L$ is called the strain. Figure 2.1 shows the typical behaviour of an elastic material when stressed. For small values of the stress,

DOI: 10.1201/9781003261773-3

BEHAVIOR OF AN ELASTIC MATERIAL

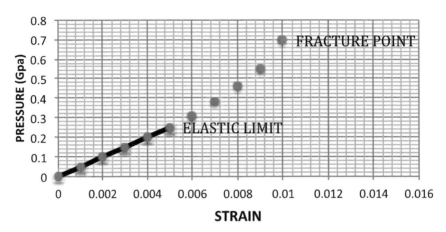

Figure 2.1 The straight line segment, whose slope is equal to Young's modulus, covers the regime where the material behaves elastically. By definition, for any strain less than or equal to the elastic limit, the material will return to its original shape (corresponding to zero strain) once the stress is withdrawn.

there is a linear relationship between the stress and the strain. The slope of this line is known as Young's modulus. If the stress is withdrawn when the stress and strain lie at any point along this linear segment, the material returns to its original shape. However, beyond a certain strain known as the "limit of elasticity", the strain-stress relationship is no longer linear. And if the stress continues to increase, eventually the fracture point is reached, where the sample breaks down.

Young's modulus E is defined by the relation:

$$E \frac{\sigma}{\left(\frac{\Delta l}{L}\right)} \tag{2.1}$$

where σ is the stress, in units of pressure (force/area), and $\Delta l/L$ is the strain. Since the strain is a dimensionless quantity, E also has units of pressure. If the stress is compressional, equation (2.1) would include a minus sign, because a reduction in the length of the body would correspond to an increase in the compressive stress.

When a material expands in one direction due to the action of a stress, the magnitude of the expansion is given by (2.1). In directions at right angles to the direction of expansion, however, the material will contract. To describe this behaviour, a second elastic parameter characteristic of the material is introduced: the Poisson ratio.

Figure 2.2 shows a sketch of a simple prismatic body, with length L, width W and height H.

If a tensile stress is applied parallel to the side marked Length, there will be an increase Δl in the value of L consistent with equation (2.1), but at the same time contractions along the other two sides. The changes Δw and Δh in the values of the height

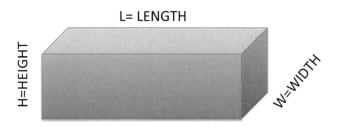

Figure 2.2 Sketch of a simple prismatic body.

and width will be negative, and will be determined by the value ν of the Poisson ratio through the strain equation:

$$\frac{\Delta w}{W} = \frac{\Delta h}{H} = -\nu \frac{\Delta l}{L} \qquad (2.2)$$

The value of the Poisson ratio is always less than 0.5.

The fact that equations (2.1) and (2.2) are linear entails that the principle of superposition is applicable to elasticity problems (Feynman et al., 1977). This principle states that, if a set of forces is applied to a body and causes certain displacements, and a second set of forces is subsequently applied to produce further displacements, the net final displacements will be the sum of the two separately. That is to say, the final result is the same as if the two sets of forces had been applied simultaneously.

Another constant of particular interest is the bulk modulus, which describes the change in volume of materials under hydrostatic pressure. If P is the confining pressure and K is the bulk modulus of the material, the volume strain, which is the quotient $\frac{\Delta V}{V}$, satisfies the equation:

$$P = -K \frac{\Delta V}{V} \qquad (2.3)$$

The parameter K (which is sometimes written also as K_b) will be later defined in differential form, as it appears in the Gassmann equation. At this stage, we will demonstrate how K can be expressed as a function of E and ν. The first step in the demonstration is to show that

$$\frac{\Delta V}{V} = \frac{\Delta l}{L} + \frac{\Delta w}{W} + \frac{\Delta h}{H} \qquad (2.4)$$

which is to say that the volume strain is equal to sum of the three linear strains.

The original volume of the body is

$$V = LWH \qquad (2.5)$$

while the volume of the body after deformation is given by:

$$V_f = (L + \Delta l)(W + \Delta w)(H + \Delta h) \qquad (2.6)$$

where the changes Δl, Δw and Δh are assumed to be small. After expansion, the right-hand side of equation (2.6) involves a sum of eight terms. All terms containing products of 2 or 3 Δs are assumed to be negligible, so we are left with just four terms:

$$V_f = LWH + HL\Delta w + HW\Delta l + LW\Delta h \tag{2.7}$$

In view of equation (2.5), it follows that $HL = V/W$, $HW = V/L$ and $LW = V/H$.
   Equation (2.7) can therefore be rewritten as

$$V_f = V\left(1 + \frac{\Delta l}{L} + \frac{\Delta w}{W} + \frac{\Delta h}{H}\right) \tag{2.8}$$

and the net change in the volume of the body is

$$\Delta V = V\left(1 + \frac{\Delta l}{L} + \frac{\Delta w}{W} + \frac{\Delta h}{H}\right) - V = V\left(\frac{\Delta l}{L} + \frac{\Delta w}{W} + \frac{\Delta h}{H}\right) \tag{2.9}$$

To determine the relationship between K, E and v, we will follow Feynman's derivation (1977), which relies on the principle of superposition.
   We first apply a compressive stress along the length of the body. The corresponding strain is:

$$\frac{\Delta l_1}{L} = -\frac{P}{E} \tag{2.10}$$

We then apply the same compressive stress to the width of the body, which results in the contraction

$$\frac{\Delta w}{W} = -\frac{P}{E} \tag{2.11}$$

However, this contraction will be accompanied by a compensating expansion along the length of the body, given by

$$\frac{\Delta l_2}{L} = -v\frac{\Delta w}{W} \tag{2.12}$$

Combining equations (2.11) and (2.12), we conclude that

$$\frac{\Delta l_2}{L} = v\frac{P}{E} \tag{2.13}$$

Similarly, if we apply the compressive stress P to the height of the body, the corresponding change in the length is given by

$$\frac{\Delta l_3}{L} = v\frac{P}{E} \tag{2.14}$$

By the principle of superposition, the total strain in the direction of the length is therefore

$$\frac{\Delta l}{L} = \frac{\Delta l_1}{L} + \frac{\Delta l_2}{L} + \frac{\Delta l_3}{L} = -\frac{P}{E}(1-2v) \qquad (2.15)$$

As the material is assumed to be isotropic, equation (2.15) applies also to the changes in the height and width. So we have

$$\frac{\Delta V}{V} = \frac{\Delta l_1}{L} + \frac{\Delta w}{W} + \frac{\Delta h}{H} = -3\frac{P}{E}(1-2v) \qquad (2.16)$$

Furthermore, from equations (2.3) and (2.16), it follows that

$$K = \frac{E}{3(1-2v)} \qquad (2.17)$$

Although many different types of elastic constants exist, the deformation of any elastic, isotropic and homogeneous body can be fully characterized by just two constants, E and v. Equation (2.17) is an example of how a modulus that was introduced for the sake of convenience (the bulk modulus) can be expressed as a function of these two parameters.

THE CONCEPT OF NORMAL AND SHEAR STRESSES

If a force is applied to a surface – such as in the example in Figure 2.3, where the surface is the upper face of a rectangular prism – the force can always be represented as the sum of components normal and tangential to the surface. These two components, when divided by the area of the top face of the rectangular prism, define the normal stress and the shear stress. If only normal stresses were to act on a body, the volume of the body would change, but the shape would not. On the other hand, if only shear stresses were to act on the body, there would be a deformation of its shape without any change in its volume.

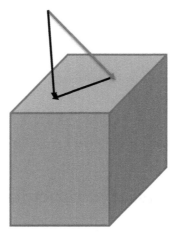

Figure 2.3 A force is applied to the top face of a rectangular prism. The force can be decomposed as the vector sum of two components: a normal and a tangential force.

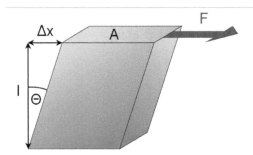

Figure 2.4 The parameters involved in the definition of the shear modulus. (From Wikipedia; article: Shear Modulus.)

THE SHEAR MODULUS

Figure 2.4 illustrates the effect of applying a tangential force to the upper face of a prism. The prism is originally rectangular, but after the application of the tangential force to its upper face, with area A, the body deforms and the prism distorts. The shear modulus μ is defined to be

$$\mu = \frac{F}{A}\frac{1}{\Delta x} \tag{2.18}$$

where the meanings of the terms on the right-hand side of equation (2.18) should be clear from Figure 2.4. The value of the shear modulus is independent of the fluid content of a body. A rock saturated with water would have the same shear modulus as an identical rock saturated with oil.

The shear modulus is one of two so-called Lamé parameters. The other (often called the "first") Lamé parameter is conventionally denoted by the symbol λ and is also sometimes used in rock physics.

This parameter is defined as

$$\lambda = \frac{vE}{(1+v)(1-2v)} \tag{2.19}$$

The shear modulus can alternatively be written as

$$\mu = \frac{3}{2}(k_b - \lambda) \tag{2.20}$$

Note here that, as mentioned above, any elastic modulus can be always expressed as a function of E and v.

RELATIONSHIP BETWEEN SEISMIC VELOCITIES AND ELASTIC MODULI

The following are probably two of the most fundamental equations of rock physics, because they relate the elastic moduli (which characterize all possible deformations of the rock) to quantities which are measured on a routine basis in oil exploration.

$$V_p = \sqrt{\frac{K_b + \frac{4}{3}\mu}{\rho_b}} \qquad (2.21)$$

$$V_s = \sqrt{\frac{\mu}{\rho_b}} \qquad (2.22)$$

In equation (2.21), the symbol V_p denotes the primary or compressional velocity of seismic waves, which is determined by the bulk modulus K_b, shear modulus μ and bulk density ρ_b of the rock. In equation (2.22), V_s denotes the shear velocity of seismic waves, which is fixed by the shear modulus and the bulk density.

The derivation of these two equations, which is not straightforward, can be found in Telford et al. (1986).

REFERENCES

Feynman, R., Leighton, R. and Sands, M., (1977) *The Feynman Lectures on Physics*, Vol. 2, Addison Wesley Publishing Company, Reading, MA, Sixth Printing.

Telford, W., Geldart, L., Sheriff, R. and Keys, D. (1986), *Applied Geophysics*, Cambridge University Press, Cambridge, Thirteenth Printing.

Chapter 3

Pore pressure review

INTRODUCTION

The pore space within the rocks is generally fully saturated with one or more fluids. These fluids are subject to a pressure called the "pore pressure", "fluid pressure" or "formation pressure". The pore pressure is the quantity that is measured in a pressure build-up test or with a formation tester (using RFT-type tools). It is also the quantity measured in a static gradient survey carried out in a well that has been shut in for a long period of time.

The pore pressure is a property that is of great interest in rock physics, from both a theoretical and a practical point of view. Consider two wells, A and B, that encounter a certain sand formation at exactly the same depth below the sea floor. If the pore pressure is much higher in well A than in well B, we can conclude that the porosity in A is also much larger than it is in B. The relationship between porosity and depth of burial is well documented, but in high-pressure zones the common maxim that "porosity decreases as the depth of burial increases" is no longer valid. This maxim is a very broad statement which will be explained in more detail below, but one principle that is almost universally true is that unexpected changes in porosity generate unexpected changes in the density and elastic properties of rocks.

Pore pressure studies are fundamental in drilling operations. The drilling mud density must always be greater than the formation pressure equivalent density (otherwise a catastrophic blowout may occur). On the other hand, the drilling mud density must always be less than the fracture pressure equivalent density to prevent the formation from breaking down and causing drilling mud losses; the fracture pressure is a quantity calculated partially from the pore pressure. The well design therefore depends critically on the results of pore pressure studies carried out prior to drilling.

Pore pressure studies are possible in clastic sections because there is an approximately linear relationship between the reciprocal of the seismic velocity and the porosity. A decrease in the velocity with depth in shale suggests a zone with a high pore pressure.

Because the pore pressure is closely related to the seismic velocities, overlaying an ordinary seismic section with the coloured distribution of seismic velocities may help to detect sealing and non-sealing faults (if they are sufficiently conspicuous), in a situation where one block has a low velocity (suggesting high pore pressure) and a neighbouring block a high velocity (suggesting normal pressures). This means that pore pressure studies can be used in a purely exploration context, though the potential of this technique is not often exploited.

DOI: 10.1201/9781003261773-4

NORMAL AND ABNORMAL PRESSURES: MOST COMMON CAUSES OF ABNORMAL PRESSURE

Figure 3.1 shows a sand aquifer, which outcrops on the left side of the drawing, fully saturated with water. A well has been drilled and is shown on the right side of the figure. The well, which encounters the sand bed in the neighbourhood of a point P at a depth D below the surface, is filled to the top with water (as a result purely of hydrostatic balance). The pore pressure P at the point P is equal to $\rho_w gD$, where ρ_w is the water density and g is the acceleration due to gravity (9.8 m/sec²). The water in the aquifer is assumed to be static.

If we calculate the pressure in kgf/cm² (where 1 kgf/cm² = 14.2233 psi), the density in g/cc and the depth in metres, the formula $P = \rho_w gD$ can be rewritten as:

$$P = \rho_w \frac{D}{10} \tag{3.1}$$

Referring to Figure 3.1, we can see that the aquifer is in contact with the atmosphere. As the water density is close to 1.0 g/cc, the observed or measured pressure should be very close to $D/10$. For example, if the depth D is 2000 m, the expected pressure should be in the order of 200 kgf/cm². If the measured pressure is close to 200 kgf/cm², this is an indication that the pore pressure is "normal". A substantial deviation from the value $D/10$ would suggest that the pressure is "abnormal" or that there is an overpressure. Implicitly, it is assumed that any rocks exhibiting a "normal" pressure are connected in some way with the atmospheric pressure at the surface, which ensures that there is at most only a slight deviation from the theoretical value of $D/10$.

We can invert equation (3.1) to solve for the density ρ:

$$\rho = \frac{10P}{D} \tag{3.2}$$

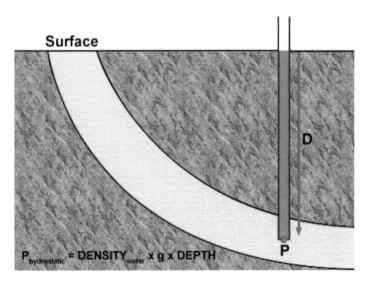

Figure 3.1 Definition of pore pressure. The case of "normal" pore pressure.

which represents the average density of the whole column of water. However, the inferred value of ρ can also be interpreted as an "equivalent density". If the value of the pore pressure at a certain depth is known, and the density estimate generated by equation (3.2) is far from 1.0 g/cc, then rather than interpreting ρ as the true density, we would treat it as an "equivalent density" whose abnormal value indicates a zone of overpressure or abnormal pressure. For example, if equation (3.2) predicts an equivalent density of 1.3 g/cc at a particular depth, it is clear that this must be due to an abnormal pressure, because it is not possible that the average density of the brine can be increased to 1.3 g/cc from the surface to the point of measurement. Table 3.1 presents a pressure assessment of this type, where the equivalent density is used to discriminate between normal and abnormal pressures. For instance, in well No. 1 (in the upper perforations), the observed pressure of 173 kgf/cm^2 was much higher than the expected pressure at the depth of interest (which is 129.7 kgf/cm^2). The equivalent density of 1.33 g/cc offers a quantitative measure of the discrepancy and implies that the formation is overpressured. For the same sand formation, the mud density used when the well is finally drilled would have to be greater than 1.33 g/cc.

The most common causes of abnormal formation pressures are: (a) substantial columns of light hydrocarbons; (b) artesian phenomena; and (c) sub-compaction of sediments (which is possibly the most common cause in the Tertiary column in the Gulf of Mexico). This list is by no means exhaustive.

Figure 3.2 demonstrates how a column of light hydrocarbons can produce an overpressure at the point marked P. Imagine a point P' lying on the hydrocarbon/water contact surface, and another point N' at exactly the same level as P', but at a point on the syncline where there is just water. The pressures at both points P' and N' are the same, that is, $P_{P'} = P_{N'}$. Let us call this common pressure P_0. Then, the pressures at the points P and N directly above P' and N' are:

$$P_P = P_0 - \rho_h gH \tag{3.3}$$

$$P_N = P_0 - \rho_w gH \tag{3.4}$$

respectively, where ρ_h in (3.3) is the hydrocarbon density and ρ_w in (3.4) is the water density. Subtracting equation (3.4) from equation (3.3) gives the overpressure

$$P_P - P_N = (\rho_w - \rho_h)gH \tag{3.5}$$

Table 3.1 Identifying normal and abnormal pressure sands in a gas field with several producing intervals, most of which are of limited lateral extent and scarce thickness

Well	Depth of reference (m)	Perforated interval (m)	Type of test	Observed pressure (kgf/cm^2)	D/10 (kgf/cm^2)	Equivalent density (g/CC)	Remarks
1	1297	1294–1300	Stat. Grad.	173	129.7	1.334	Overpress
1	1803	1803–18041)4	Stat. Grad.	248	180.3	1.375	Overpress
2	1753	1735–1740	Stat. Grad.	320	175.3	1.825	Overpress
3	595	595–634	DST {open h.	56.3	59.5	0.946	Normal

Surface

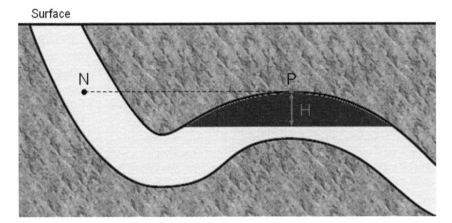

Figure 3.2 An example of overpressure produced by a column of hydrocarbon with a thickness equal to H. In this case, despite the overpressure, the aquifer is in contact with the atmosphere.

It is evident from this elementary analysis that the pressure at the point P, on the top of the hydrocarbon reservoir, is greater than the pressure at the point N, which is at the same elevation as P, but sits in the middle of the aquifer. The greater the contrast between the density of the water and the density of the hydrocarbons, and the greater the height of the hydrocarbon column, the greater will be the excess pressure at the point P. If the point N is itself at a normal pressure, in the trap, the pressure will also be normal at the points of hydrocarbon/water contact.

An artesian phenomenon generally occurs when an aquifer outcrops at one point at a relatively high elevation and lies at a substantial distance below the surface at others. Figure 3.3 gives a conceptual illustration of this situation.

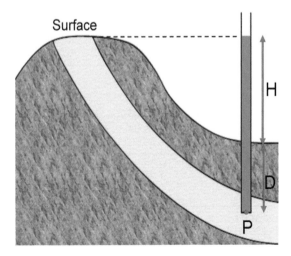

Figure 3.3 Conceptual illustration of the geological conditions leading to an artesian phenomenon.

Assume here that the surface at the location of the well coincides with sea level. If the pore pressure P at the point P in the well was normal, it would be given by the usual formula $P = \rho_w\, gD$, where D is the depth of P below the surface. The water level in the well would then just reach the top of the well at the surface. However, in the situation shown the water level in the well would continue to rise much higher before equilibrating with the atmospheric pressure, by an extra distance equal to H, which is the difference in elevation between the outcrop and the surface at the well. In practice this would mean that the well would be subject to a natural outflow of water, once it is opened. The excess pressure at the point P would be equal to $\rho_w\, gH$. The analysis of the problem here is somewhat oversimplified, because in a realistic situation the water in the aquifer would most likely be in motion.

It seems probable that the main cause of overpressures in the Tertiary column in the Gulf of Mexico is the sub-compaction of sediments. As new sediments are deposited, the sediments underlying them are slowly compressed by the weight of the newest sediments. Under normal circumstances, the effect of this compaction is to expel water from the pore space in the older sediments, leading to a decrease in porosity.

The conventional expectation, therefore, is that the porosity of both sands and shale should decrease with the depth of burial. However, if for some reason the water cannot escape (if, for instance, the water-bearing sediments are surrounded by sediments that are extremely impervious), the porosity of the rocks being compacted cannot reduce, and the fluid in the pores will carry a higher proportion of the load (*i.e.* the fluid is subject to an overpressure).

The somewhat vague relationship between porosity and pore pressure plays a fundamental role in the identification of high-pressure zones using seismic velocities (and also in wells with sonic logs). Normally, the shallowest rocks in a certain area are in contact, in one way or another, with the atmosphere and have a "normal" pressure, so their porosity decreases with depth quite regularly. This reduction in porosity in turn causes an increase in the p-wave velocity, and a "normal compaction trend" can be identified in seismic surveys. In many instances in the Gulf of Mexico, it is possible to probe thousands of metres of Tertiary sediments without seeing any departure from the normal compaction trend, as all the materials in the column are inferred to be at normal pressure. However, if a zone of high pressure is encountered, the normal compaction trend breaks down. An increase in porosity results in a reduction of the seismic velocity. In practice, this type of diagnosis is most reliable when applied to shale, but with seismic data alone we cannot generally separate sand from shale beforehand. In any case, a reduction of the seismic velocity or deviation from the normal compaction trend is normally regarded as evidence of a high-pressure zone. In fact, what really controls the porosity is a property called the net overburden pressure (NOBP), which will be defined shortly, but for all practical purposes the considerations described here are reliable indicators.

OVERBURDEN PRESSURE AND NET OVERBURDEN PRESSURE

The overburden pressure OP at a certain depth in a formation is the pressure due to the weight per unit area of all the overlying material (solid and fluid). Mathematically, this can be expressed as:

$$OP = OP_0 + \frac{z}{10}\, \rho_{AVER} \tag{3.6}$$

where the overburden pressure is measured in kgf/cm^2, OP_0 is the overburden pressure at the depth $z = 0$ and z itself is the depth below the sea floor. The depth $z = 0$ marks the location of the proper sea floor, which means that OP_0 is equal to the pressure exerted by the column of water extending from sea level down to the sea floor (as the atmospheric pressure is ignored in sub-ocean drilling). ρ_{AVER} in equation (3.6) is the effective density of all the materials above the point where the overburden pressure is being measured. Because the density typically varies with depth, ρ_{AVER} is in fact the average density of the rocks and fluid above the depth z. The identification of ρ_{AVER} with the average density can be demonstrated more formally, as follows.

Suppose that N measurements of the density have been taken in a vertical section, from the sea floor down to a depth equal to z. If the measurements are equally spaced, and are separated by a distance Δh, (the symbol Δ remains) then the number of measurement points in the column is:

$$N = \frac{z}{\Delta h} + 1. \tag{3.7}$$

If the overburden pressure OP is integrated numerically from the first to last measurement points in the section, with the two end points given a half weight, then:

$$OP_N = OP_0 + \frac{\Delta h}{20}(\rho_0 + \rho_N) + \frac{\Delta h}{10}\sum_{k=2}^{N-1}\rho_k = OP_0 + \frac{\Delta h}{10}\rho_{AVER}(N-1) \tag{3.8}$$

where by definition the average density in the column is

$$\rho_{AVER} = \frac{\Delta h}{2(N-1)}(\rho_0 + \rho_N) + \frac{\Delta h}{N-1}\sum_{k=2}^{N-1}\rho_k. \tag{3.9}$$

Combining equations (3.7) and (3.8) then gives

$$OP_N = OP_0 + \frac{z}{10}\rho_{AVER} \tag{3.10}$$

It has been noticed that, in a large number of wells in the Tertiary section of the Gulf of Mexico, the overburden pressure tends to increase as a quadratic function of the depth z. That is, $OP = az^2 + bz + c$ for some set of constants a, b and c. This behaviour is so common as to be almost universal. If the overburden pressure is regarded as a continuous function of the depth, we can solve equation (6) for the average density as a function of z. This gives:

$$\rho_{AVER} = 10\left(\frac{az^2 + bz + c - OP_0}{z}\right). \tag{3.11}$$

Furthermore, given that OP should reduce to OP_0 at $z = 0$, it follows that $c = OP_0$, and equation (3.11) becomes:

$$\rho_{AVER} = 10(az + b). \tag{3.12}$$

On the other hand, if we differentiate OP with respect to z, and identify the resulting expression with $\rho / 10$, where ρ is the local value of the density in the column at depth z, we get:

$$\frac{dOP}{dz} = 2az + b = \frac{\rho}{10}. \tag{3.13}$$

That is,

$$\rho = 10(2az + b). \tag{3.14}$$

That is, the local density ρ is a linear function of the depth.

In many cases, the coefficient a in equation (3.13) is so small that it can be ignored, and the gradient dOP / dz is a constant, equal to 1 psi/ft. A preliminary estimate of the overburden pressure can therefore be made by assuming a constant gradient of 1 psi/ft, which is the value generally accepted as the average gradient of the overburden pressure in the Gulf of Mexico. (And similar values are widely reported from other parts of the world.)

If the well information is available, we can instead estimate the overburden pressure by numerically integrating the density log. If at depth equal to Z_1 (the first depth at which we have density log data), the overburden pressure is OP_1, subsequent values of the overburden pressure can be calculated from the formulas:

$$OP_2 = OP_1 + (Z_2 - Z_1)\frac{\rho_2 + \rho_1}{2}g$$

$$OP_n = OP_{n-1} + (Z_n - Z_{n-1})\frac{\rho_n + \rho_{n-1}}{2}g$$

If the pressures are expressed in kgf/cm^2, the depths in metres and the densities in g/cc, these equations become:

$$OP_n = OP_{n-1} + \left(\frac{Z_n - Z_{n-1}}{10}\right)\left(\frac{\rho_n + \rho_{n-1}}{2}\right) \tag{3.15}$$

However, it is usually the case that the density log is missing from the first 500 m or so below the mudline. To circumvent this problem, we can proceed as follows:

a. To the depth below the mudline where the first valid density value has been recorded, assign (arbitrarily) an overburden pressure of 0.
b. Then apply equation (3.15) to calculate all the succeeding values of the overburden pressure. If these numbers, which represent an arbitrary overburden pressure, are plotted against the depth below the mudline, the resulting curve will be parallel to the real overburden pressure (see Figure 3.4), as the derivative with respect to depth of the real curve (if it is parabolic) will match the derivative of the arbitrary curve.
c. Fit the observed data with a quadratic function of the form $OP = az^2 + bz + c$. (However, only the constants a and b are important here. The constant c will be discarded.)

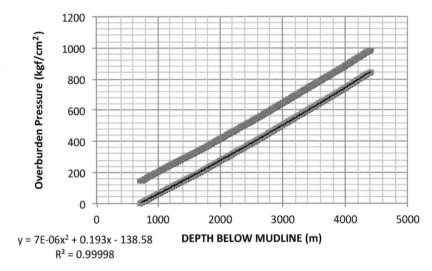

$y = 7E\text{-}06x^2 + 0.193x - 138.58$
$R^2 = 0.99998$

Figure 3.4 The lower curve represents the "temporary" values of the overburden pressure. Note that the curve is almost a perfect parabola (its equation and the square of the correlation coefficient can be seen at the bottom left). We use this parabola to calculate the overburden pressure at 705.3 m, given that the value of the overburden pressure at $z = 0$ is known. Once the value of the overburden pressure at 705.3 m is known, we can calculate the upper, final curve.

d. We know that at $z = 0$ (at the mudline), the overburden pressure is equal to the weight of the column of water extending from the sea level to the mudline (meaning that OP_0 is simply $\rho_w\, gD$, where the water density is the density of seawater and D is the depth of the mudline below sea level). Replacing c with the calculated value OP_0 then gives $OP = az^2 + bz + OP_0$, which is now the true equation of the overburden pressure as a function of depth. We can use this equation to calculate the overburden pressure at the depth where the first valid density measurement was reported (to which we previously assigned a value of 0).

e. Add this value to the set of data points to obtain an estimate of the true overburden pressure as a function of depth.

As a practical demonstration of this procedure, we now summarize the results of applying the method to data from a real well. The depth of the mudline was 19 m below the sea level. The well had its first valid density reading at 705.3 mbml and the last measurement was taken at 4404 mbml.

Table 3.2 shows a small fraction of the calculations that were performed. The first column lists the depth in metres below the mudline, followed by the observed density, the original values of OP calculated from equation (3.15) on the assumption that $OP = 0$ at 705.3 mbml (and then used to fix the parameters in the equation for the parabola), the final overburden pressure once the overburden pressure of the shallow section has been added, and the overburden pressure gradient, which does actually vary along the column, although the variation cannot be appreciated in data taken from such a short interval. Both pressure columns are measured in kgf/cm^2, but the gradient is in psi/ft.

Table 3.2 An extract from the overburden pressure calculations over a short depth interval.

Depth (bml)	Rhob	Op (temp.)	Op. (Final)	Grad. Op
705.3204	2.2483	0.0000	141.5414	0.8794
705.4728	2.3046	0.0347	141.5761	0.8794
705.6252	2.3167	0.0699	141611.3	0.8794
705.7116	2.3309	0.1053	141.6467	0.8794
705.9300	2.3509	0.1410	141.6824	0.8794
706.0824	2.3646	0.1769	141.7183	0.8794
706.2348	2.3655	0.2130	141.7544	0.8794
706.3872	2.3406	0.2488	141.7902	0.8794
706.5396	2.2776	0.2840	141.8254	0.8794
706.6920	2.1974	0.3181	141.8595	0.8794
706.8444	2.1403	0.3512	141.8926	0.8794
706.9968	2.1208	0.3836	141.9250	0.8795
707.1492	2.1281	0.4160	141.9574	0.8795

Depths are in metres, densities in g/cc, pressures in kgf/cm^2 and the gradient of the overburden is in psi/ft.

Incidentally, the average value of the overburden pressure gradient between the two logged depths at 705.5 and 4404 mbml was calculated to be 0.99 psi/ft, while the average of the same variable between mudline and 4404 mbml was 0.97 psi/ft. These are both very close to the standard value of 1 psi/ft reported in the literature.

An alternative way to estimate the overburden pressure at shallow depths is to use Miller's method (Zhang, 2013). This method is in fact very popular among rock physicists. From the examination of a large amount of data on the porosity of shallow sediments, an empirical equation relating the porosity to the depth has been constructed, of the form:

$$\phi = \phi_a + \phi_b \exp(-KD^N), \tag{3.16}$$

where D is the depth below the mudline in feet, and the other parameters are constants. The optimum values for these constants, in the Gulf of Mexico, are $\phi_a = 0.35$, $\phi_b = 0.35$, $K = -0.0035$ and $N = 1/1.09$.

Miller's method can be applied to the data set described above as follows. Once the porosity ϕ has been calculated from the mudline to 705.3 m, the corresponding densities can be calculated using the formula:

$$\rho_b = \rho_{ma}(1-\phi) + \rho_f \phi, \tag{3.17}$$

where the matrix density ρ_{ma} is assumed to be 2.68 g/cc and fluid density ρ_f to be 1.03 g/cc.

Once the values of the density have been calculated for depths down to the depth of interest, they can be integrated using equation (3.15) to estimate the overburden pressure. What really matters in the case studied here is the overburden pressure at the depth where the first density log value was reported.

Table 3.3 Using Miller's method to estimate the overburden pressure at 705.5 mbml, where the first value of the density log is recorded.

Mudline (m)	Mudline (ft)	Phi (Miller)	Rhob	Op
700.00	2296.59	0.3550	2.0907	140.2549
700.50	2298.23	0.3550	2.0907	140.3595
701.00	2299.87	0.3550	2.0907	140.4640
701.50	2301.51	0.3550	2.0907	140.5686
702.00	2303.15	0.3550	2.0907	140.6731
702.50	2304.79	0.3550	2.0908	140.7776
703.00	2306.43	0.3549	2.0908	140.8822
703.50	2308.07	0.3549	2.0908	140.9867
704.00	2309.71	0.3549	2.0908	14.0912
704.50	2311.35	0.3549	2.0909	141.1958
705.00	2312.99	0.3549	2.0909	141.3003
705.50	2314.63	0.3549	2.0909	141.4049

Note that the overburden pressure at 705.5 mbml is 141.4 kgf/cm^2. Using the method extracted in Table 3.2, the result is 141.54 kgf/cm^2 at 705.3 mbml, which is virtually the same.

An extract from the calculations is presented in Table 3.3. It is interesting to note that both methods yield almost identical results. From Table 3.2, we can see that at a depth of 705.3 mbml the overburden pressure calculated using the first method is 141.54 kgf/cm^2, while Miller's method according to Table 3.3 yields a value of 141.4 kgf/cm^2 at 705.5 mbml.

When there are no previous wells in the area, the densities can be calculated from the interval seismic velocities using Gardner's formula:

$$\rho_b = 1.741 V_P^{0.25}, \tag{3.18}$$

where the compressional velocity V_P is measured in km/sec and the density ρ_b in g/cc.

The Net Overburden Pressure is defined as the difference between the overburden pressure and the pore pressure. It is an important quantity, because the porosity of the rocks ultimately depends on it – that is, on the difference between the overburden and the pore pressures – but not on these quantities separately.

In Chapter 4, it will be shown that the porosity of an *elastic* material depends on the difference between the overburden pressure and the pore pressure. However, compacted sedimentary materials could not plausibly be described as "elastic". It is reasonable to treat rocks as elastic materials when a seismic wave passes through them. The strains are infinitesimal and the rock returns to its original condition once the disturbance dies away. However, the stresses and corresponding strains involved in the compaction of rocks are substantial. It is difficult to believe that a compacted rock would return to its original condition if the stresses were removed. Nonetheless, it is generally accepted that an increase in the Net Overburden Pressure results in a decrease in the porosity and vice versa, even though the rock itself may not be elastic. This theoretical relationship is implicitly or explicitly assumed in both the Eaton and Bowers formulae for estimating pore pressure, as will be seen later.

In formations where the pressure is normal, the NOBP and the depth are directly proportional and so, in such situations, it is possible to infer a correlation between the porosity and the depth.

Returning to the example mentioned in the Introduction to this chapter, consider two wells, A and B, that encounter an oil sand at 1000 m below the sea floor. Both have approximately the same overburden pressure. However, the pressure in well A is 2000 kgf/cm^2, while the pressure in well B is 1550 kgf/cm^2. Which of the sands is more porous? (Assuming that the water depth is the same at both locations.)

The NOBP at A is: $NOBP_A = (OP_0 + \rho_{1000}/10) - (2000)$ and the NOBP at B is: $NOBP_B = (OP_0 + \rho_{1000}/10) - (1550)$.

Given that the water depth is the same at both locations, subtracting the two NOBPs gives:

$$NOBP_B - NOBP_A = 450 \text{ kgf/cm}^2$$

That is, the NOBP is 450 kgf/cm^2 higher in well B than in well A. According to the conventional hypothesis, the porosity should therefore be greater in well A, because the NOBP is smaller there.

In connection with this, it should be noted that if two wells encounter sands at the same depth below the sea floor, but the water depth is 200 m in one of them and 2000 m in the other, the porosities of the sands will still be approximately the same, provided they are both at normal pressure. The reason for this is that in both cases the NOBP is the same. We can model the pore pressure using the formula $P = P_0 + \rho_w Z/10$, where Z is the depth below the sea floor and P_0 is the pore pressure at the depth of the sea floor (and so is the same as OP_0). These two quantities cancel one another when the NOBP is calculated. In principle, therefore, the water depth does not affect the porosity of sediments.

THE GLUYAS-CADE CORRELATION OF POROSITY VS. DEPTH FOR CLEAN, UNCEMENTED SANDS

The Gluyas-Cade (1997) correlation is also an important tool, because it predicts the porosity in clean, uncemented quartz sands as a function of depth if the sands are normally pressured. Modifications to the formula can also be made to account for abnormally high pressures. This correlation serves as an envelope for the values of the porosity that might be expected in a vertical section prior to drilling. In principle, clean, uncemented sands have the maximum possible porosity. As soon as cement or clayey materials are added to the matrix, the porosity is reduced. In other words, the porosity of sand as a function of depth is expected to always be equal to or less than the values provided by the Gluyas-Cade correlation.

The correlation, for normally pressured sands, is:

$$\phi = 0.5 \exp\left(\frac{-10^{-3} z}{2.4 + 0.0005z}\right), \tag{3.19}$$

where z is the depth in metres and ϕ is the porosity. According to its authors, the accuracy of this correlation is ±0.025, or 2.5 porosity units.

In this case there is no ambiguity: the effective and total porosity of the sands are exactly the same, due to the absence of clayey materials.

In the case of an overpressure, the authors provide a formula for calculating an "equivalent depth", which needs to be introduced into equation (3.19) to generate the final value of the porosity. The equation for the equivalent depth is as follows:

$$z' = z - \frac{u}{(\rho_0 - \rho_w)(1-\phi_{AVER})g},$$
(3.20)

where u is the excess pressure in Pa, ρ_0 is the grain density in kg/m^3, ρ_w is the water density (in the same units), g is the acceleration due to gravity in m/sec^2 and ϕ_{AVER} is the average porosity down to the depth z. Typical values are 2650 kg/m^3 for the grain density and 0.2 for the average porosity.

Figure 3.5 shows the typical range in the porosity of sands with varying degrees of shale content, as a function of depth in a normally pressured well. The upper curve is the Gluyas-Cade correlation, and as expected, it forms an approximate upper envelope

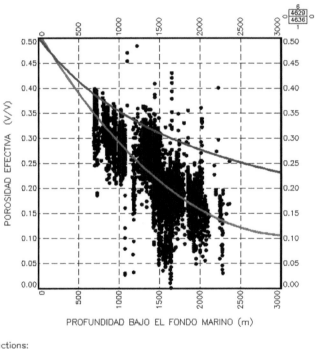

PROFUNDIDAD BAJO EL FONDO MARINO (m)

Functions:
dummy: Regression Logs: MUDLINE, EVAL.PHIE, CC: 0.735831
        PHIE = (0.509381 − 0.000258988*(MUDLINE) +
        4.15726e−08*(MUDLINE)**2)
gluyas: Regression Logs: MUDLINE, EVAL.PHIE
        PHIE =
        (0.5*exp((−0.001)*(mudline)/(0.0005*(MUDLINE)+2.4)))

Figure 3.5 The effective porosity of sands (containing up to 30% shale) as a function of depth in a normally pressured section. The upper line represents the Gluyas-Cade correlation, which as expected shows generally higher porosities than were actually measured in the (less than clean) sands. The lower line is the least-squares fit to the data. The dispersion visible in the cloud of points is due to the fact that several other factors (not just the NOBP) control the porosity of the rocks.

for the porosity data. Most of the data points lie below the Gluyas-Cade line, which is applicable only to clean, uncemented sands.

At this point, equations (3.19) and (3.20) will be used to make some calculations to estimate the effects of pore pressure on porosity.

Figure 3.6 shows the effect of overpressure on the porosity of clean, uncemented sands. The pressures range from normal (on the lowest curve) through equivalent densities of the fluid equal to 1.2, 1.4 and 1.6 g/cc on each succeeding curve. The equivalent density is calculated from the formula $\rho_{eq} = 10\ PP/z$, where PP is the pore pressure recorded at z metres below the mudline. An equivalent density of 1.6 g/cc implies that the pore pressure is about 60% greater than the normal or hydrostatic pressure. This is a moderately high pressure. The point to note is that, as the pore pressure increases, considerable changes in the porosity of the sand can occur. In practice, the porosity depends on the NOBP. If, at a certain depth, the overburden pressure remains constant but there is an increase in the pore pressure, the NOBP will fall and hence the porosity will increase.

The central message of this section has been to highlight the important role that the pore pressure plays in the control of the rock porosity in general, despite the fact that the curves in Figure 3.6 are applicable only to clean, uncemented sandstones, which is a somewhat idealized rock type.

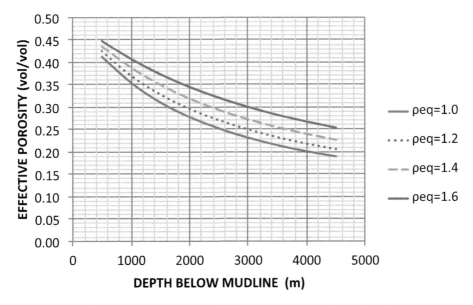

Figure 3.6 The effect on the porosity of clean, uncemented sands of an excess pore pressure (*i.e.* one exceeding the hydrostatic pressure). The lowest line corresponds to the case of normal pressure. It is apparent that the effect of an overpressure on the porosity can be considerable. The pore pressure is characterized here by the "equivalent pore pressure". An equivalent density of 1.2, for instance, implies that the pressure is 20% greater than the corresponding normal or hydrostatic pressure.

CALCULATION OF THE PORE PRESSURE: THE EATON AND BOWERS EQUATIONS

When discussing the role of the sub-compaction of sediments in causing overpressure, we stated that, in normally pressured zones, the porosity decreases with depth. In a zone with normal pressure, there will be a clear increase in the seismic velocity (or the inverse of the sonic log, in the case of a well) with depth. This regular increment of the velocity with depth defines a "normal compaction trend line" (NCTL). The ideal lithology in which this type of trend can be observed is shale. Because argillaceous sediments have less variation in texture than sandstones, their porosity is more directly controlled by the Net Overburden Pressure.

When dealing with seismic data, it is very difficult to separate sand from shale. However, if an abnormal pressure zone is encountered, the seismic velocity across it will be lower than predicted by the NCTL. The difference between the velocity predicted by the NCTL and the actual velocity forms the basis for quantitative estimation of the pore pressure, either explicitly or implicitly.

There are several equations that are commonly used to estimate pore pressure. We will discuss just two of them: the Eaton and the Bowers equations.

THE EATON EQUATION READS

$$PP = OP - (OP - HP)\left(\frac{V_{OBS}}{V_{NCTL}}\right)^3 , \tag{3.21}$$

where PP is the calculated pore pressure, OP is the overburden pressure, HP is the hydrostatic or normal pressure, V_{OBS} is the observed velocity and V_{NCTL} is the velocity predicted at that depth by the NCTL. If the zone follows the NCTL, the two velocities are the same and the pore pressure PP is identical to the hydrostatic or normal pressure HP.

Figure 3.7 shows a very well-defined NCTL over a section of almost 3000 m. However, the bottom of the well might be in an overpressured zone. (It is known that there were no "kicks" at the bottom of this well, when drilling with the same mud as a few metres above it. This suggests, but does not prove, that the pressure at the bottom is still normal.)

It is evident that, when using the Eaton equation (3.21), it would be convenient to have an analytic formula for the NCTL (describing it, for example, as a straight line, or a logarithmic or exponential line). In the example shown in Figure 3.7, this is not a problem because almost any line will fit the data equally well, and extrapolation to deeper depths would probably give similar results. However, there are other situations in which the construction of the line is not so straightforward.

Figure 3.8 plots a set of seismic data (with the seismic interval velocity on the vertical axis) for which it is difficult to define a unique NCTL.

Note that three different curves (exponential, linear and a function of the square root of the depth) fit the data for the shallow depths equally well. However, at greater depths the values of the quotient V_{OBS}/V_{NCTL} will be very different, depending on the line chosen. Equation (3.21) will then predict very different values for the pore

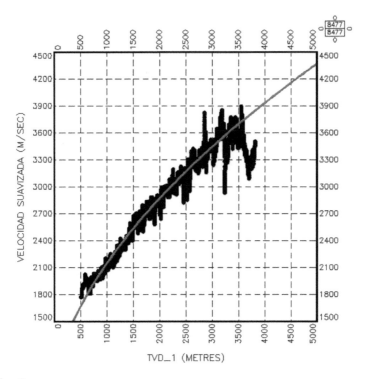

Functions:
vtrend: Regression Logs: REFERENCE.TVD, VSMOOTH**2, CC: 0.979721
      VSMOOTH =
      SQRT(984207.6 + 3623.406*(TVD))

Figure 3.7 Example of a very well-defined NCTL over 3000 m of data, indicating that the well is at normal pressure. However, there is a hint of an overpressured zone right at the bottom the well.

pressure. It is therefore necessary to check if there are any independent constraints on the shape of the NCTL.

Hottmann and Johnson (1965) present a plot of depth against slowness for a substantial number of normally pressured shales. The data correspond to Miocene to Oligocene shales from Upper Texas and the southern Louisiana Gulf coast. The x axis, corresponding to the slowness Δt, is scaled logarithmically, while the y axis, representing the depth z, is linear. The data points, taken from depths ranging from 2000 to 14000 ft, conform to an almost perfect straight line. This implies a relationship of the form:

$$ln(\Delta t) = Az + B \tag{3.22}$$

Equation (3.22) can alternatively be written as:

$$\frac{1}{V} = e^{Az+B}, \tag{3.23}$$

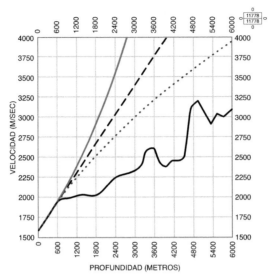

Functions:
anoma_trend_logaritmico: Regression Logs: REFERENCE.DEPTH, SEISMIC_ULTIMO.TRC_1, CC:
0.999553
$y = 10^{**}(3.20328 + 0.000144803 \cdot (x))$
anoma_trend_linear   : Regression Logs: REFERENCE.DEPTH, SEISMIC_ULTIMO.TRC_1, CC:
0.999108
$y = (1585.383 + 0.604187 \cdot (x))$
anoma_trend_short   : Regression Logs: REFERENCE.DEPTH. SEISMIC_ULTIMO.TRC_1**2. CC:
0.998239
$y = sqrt(2470112 + 2190\ 815 \cdot (x))$

Figure 3.8 There is a sharp reduction in the seismic velocity at a very shallow depth in this
example, implying that data points lying on the NCTL are scarce. It is assumed
here that the overpressured zone begins at the point where the velocity levels
off at about 500 mbsf. Different NCT lines fit the data equally well.

which (given that A is negative) entails that the velocity increases exponentially with
depth. That is to say, equation (3.23) can be written in the form $V = p \exp(qz)$, where p
and q are positive constants.

Another piece of empirical information is that, in a great number of wells in the
Gulf of Mexico (and probably elsewhere), a plot of depth against cumulative time
(which is calculated by integrating the sonic log) shows an almost perfect quadratic
relationship, of the form:

$$z = at^2 + bt + c, \tag{3.24}$$

where t is the cumulative time and a, b and c are constants. Figure 3.9 shows data from
two wells where the relation between these variables is quadratic. Note that this relationship occurs only when the well is normally pressured.

Now, the derivative of equation (3.24) reads:

$$\frac{dz}{dt} = V = 2at + b. \tag{3.25}$$

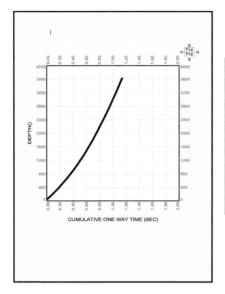
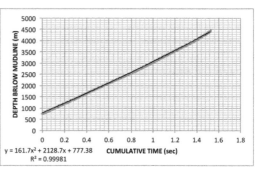

Figure 3.9 Depth vs. cumulative time (integrated from the sonic log) for two very different wells. The plot on the left corresponds to an onshore well. The other plot corresponds to an offshore well. The relationship is clearly quadratic for the onshore well, and it is also quadratic for the offshore well, although this is not so obvious. Both wells are normally pressured.

On solving equation (3.24) for t and substituting the result into equation (3.25), it follows after some manipulation that:

$$V = \sqrt{b^2 - 4a(c - z)}. \tag{3.26}$$

Equivalently, equation (3.26) can be written as:

$$V = \sqrt{B + Az}, \tag{3.27}$$

where A and B are arbitrary constants. Equation (3.27) entails that, in zones of normal pressure, the velocity is the square root of a linear function of the depth.

The principal difference between the exponential NCTL claimed by Hottmann and Johnson, and the square root function in equation (3.27), is that the derivative dV/dz increases with depth if equation (3.23) applies and decreases with depth if equation (3.27) applies. In other words, the velocity associated with the exponential NCTL (equation 3.23) grows more rapidly as the depth increases, whereas the velocity associated with the square root NCTL (equation 3.27) increases less rapidly with depth. A difference of this type can be detected visually by inspection of a plot of velocity against depth. Depending on the features of the trend, either type of function could be used to build the NCTL. However, as in situations like that shown in Figure 3.8, it is possible that the interval at normal pressure might be too short to fix a unique function.

The identification of the correct NCTL is probably the most serious obstacle in the calculation of the pore pressure using either the Eaton or Bowers method.

THE BOWERS FORMULA

The Bowers formula stipulates that the Net Overburden Pressure (the overburden pressure minus the pore pressure, as before) is related to the seismic velocity V in shale through the equation:

$$\sigma = a\left(V - V_0\right)^b,\tag{3.28}$$

where σ is the NOBP, V_0 is usually taken to be 1524 m/s and a and b are fitting parameters.

As a first step – as in Eaton's method – it is necessary to identify the part of the vertical section that is at normal pressure. Once this interval has been identified, the NOBP is calculated at all the points in the interval. This can be done because the overburden pressure is known beforehand, as is the pore pressure (the points are assumed by fiat to be at normal pressure).

Figure 3.10 is a log-log plot, with V-1524 on the x axis and the NOBP on the y axis. According to equation (3.28) the cloud of points should lie on a straight line in a log-log

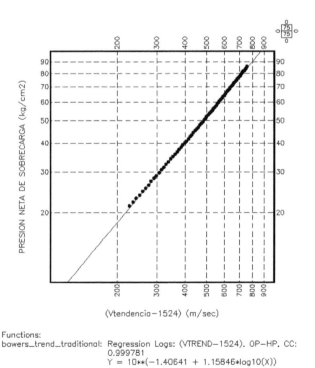

(Vtendencia−1524) (m/sec)

Functions:
bowers_trend_traditional: Regression Logs: (VTREND−1524), OP−HP, CC: 0.999781
Y = 10**(−1.40641 + 1.15846*log10(X))

Figure 3.10 An illustration of the Bowers method. The x axis represents the velocity of normally pressured points and the y axis corresponds to the NOBP. From this plot, the two parameters a and b can be calculated.

plot, and this is true for Figure 3.10. The constants a and b are the best-fit parameters for the straight line in the figure.

To calculate the pore pressure at all the points where there is an abnormal pressure, we use equation (3.28) in conjunction with the definition of the NOBP, so that:

$$PP = OP - a\left(V - V_0\right)^b . \tag{3.32}$$

This is essentially the Bowers method. The results obtained by the Bowers and Eaton methods are generally similar, and both produce reasonable matches, in intervals with abnormal pressure, with the pressure measured directly using RFT (Repeat Formation Tester) and other techniques.

Another way to check the accuracy of these methods is to compare the pressures calculated from them with the mud weight (in wells that have already been drilled, of course). If a well was drilled normally over a certain interval, the equivalent density is calculated from the pore pressure using the formula $\rho_{eq} = 10 \; PP/z$, where z is now the depth below the rotary table, and must be less than the mud density (otherwise a blowout might occur). Again, both methods produce acceptable results.

LITHOLOGICAL PROBLEMS

In a vertical section containing a thick column of shale at normal pressure, it is quite simple to establish an NCTL. These conditions are generally the norm rather than the exception. The Tertiary sedimentary column in the Gulf of Mexico consists predominantly of shale. (This is also true elsewhere. Data from several authors around the world, compiled by Pettijohn et al. (1972), suggest that the proportion of shale in sediments is much greater than sandstone or limestone.) Furthermore, although it is commonly encountered in wells, overpressure is typically not a feature of very shallow sediments. Hence, in most circumstances it is possible to define an NCTL.

In a zone where there is a relative abundance of sandstone or limestone, the NCTL may be more difficult to determine, and – even if an NCTL has been found – the problem is compounded if these two lithologies are also abundant in the overpressured zone. As a consequence, errors in the calculation of the pressures may occur.

Figure 3.11 gives an example of the problems extraneous lithologies can cause in the calculation of pore pressure. Down to about 1000 mbRT, the velocity points seem to follow a clear trend, and an NCTL can be established. All the velocity points lying below the NCTL are, in principle, overpressured. However, at about 3900 mbRT there is a spike in the velocity. Because the data is taken from a well, it is known that this increase in velocity is due to the presence of a carbonate body. But the existence of the body was unknown prior to drilling. If we were dealing with seismic data, it would be easy to conclude after smoothing the velocity data that the column is normally pressured everywhere, which in turn could lead to catastrophic results if this inference was used as a guide for drilling. Incidentally, the eventual increase in velocity below about 5200 mbRT corresponds to the top of the Cretaceous, where pore pressure calculations are no longer valid. It is evident from Figure 3.11 that below the carbonate body there is an overpressured zone, which in fact extends from 1000 to 5200 mbRT.

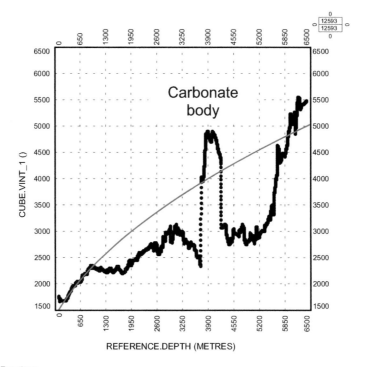

Functions:
vtrend_kinil : Regression Logs: DEPTH, CUBE.VINT**2, CC: 0.986378
VINT = SQRT(1966785 + 3592.905*(DEPTH))

Figure 3.11 The zone of high velocity at about 3900 mBRT is due to the presence of a carbonate body. If the lithology were unknown, the pore pressure would be calculated incorrectly.

PORE PRESSURE CALCULATIONS IN LIMESTONES, AND THE DIFFICULTY OF DOING THIS WITH VELOCITY DATA ALONE

Clastic sediments, such as sandstones and shales, are granular bodies in which the individual grains are in contact with their nearest neighbours, leaving a continuous three-dimensional network of voids between them. This intergranular space accounts for the porosity and permeability of sandstones. In the case of shale, this continuous three-dimensional space still exists, but the small sizes of the grains constrain the diameters of the pores to be equally small, which makes shales very impermeable. Whatever the size of the pores, there is an empirical equation which links the porosity with the slowness or transit time (which is the inverse of the seismic velocity). This equation is the Wyllie equation, which will be examined and justified in later chapters. However, at this stage, it is sufficient to note that the equation works well with intergranular materials. The Wyllie equation reads:

$$\Delta t = \Delta t_{ma} + \left(\Delta t_f - \Delta t_{ma} \right)\emptyset, \tag{3.30}$$

where Δt is the measured transit time, Δt_{ma} is the transit time in the matrix or solid part of the rock (which is 50–55 μs/ft for sandstones and has larger values for shale), Δt_f is the inverse of the fluid velocity (about 189 μs/ft for water) and ϕ is the porosity of the rock. Broadly speaking, this means that there is an approximately linear relationship between the transit time and the porosity in clastic materials.

However, only some limestones have an intergranular texture and are correctly described by the Wyllie equation. In many other limestones the p-wave velocity depends on the value of the porosity in other, more complicated ways. It appears that the shape of the pores has a strong control over the relationship between the seismic velocity and the porosity. (This topic will be discussed in more detail in Chapter 9.)

The conclusion to be drawn from this is that a change in the slowness or the velocity inside a carbonate formation may be due not to a change in the porosity itself, but rather to a change in the shape of the pores. Figure 3.12 illustrates this problem.

In Figure 3.12, which plots data from an actual carbonate formation, it is evident that the relationship between the velocity and porosity is poorly determined. It is not really possible to explain the various velocity measurements in terms of porosity changes, and for this reason it is extremely difficult to use seismic data to estimate the pore pressure in carbonates. Note that the porosity values in Figure 3.12 were calculated using Neutron and Density logs.

There is yet another reason why it is more difficult to detect overpressured zones in carbonates, which is that they are "stiffer" than shale. For a given change in the NOBP, the change in the porosity of a carbonate will be much smaller than the corresponding change in the porosity of a shale.

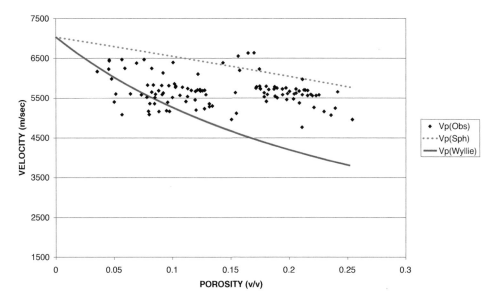

Figure 3.12 Ambiguity in the relationship between the velocity and porosity in carbonates. The upper line represents the relationship between the velocity and porosity in the case of spherical pores, which is characteristic of vuggy carbonates with unconnected pores (*cf.* Chapter 9). The lower line is the Wyllie equation. The rock is a pure limestone, devoid of clay and dolomite.

The following equation, which follows from material in Chapter 4 combined with the Krief equation (*cf.* Chapter 6), expresses the concept of stiffness more quantitatively. If it is assumed that the rocks are elastic, then the derivative of the porosity ϕ with respect to the NOBP σ is:

$$\frac{d\varnothing}{d\sigma} = \frac{1}{K_0}\left[1-(1-\varnothing)^{1-3/(1-\varnothing)}\right],\tag{3.31}$$

where K_0 is the incompressibility of the solid part of the rock.

The incompressibility of a shale may be in the order of 20 GPa, whereas the incompressibility of the solid part of a limestone is about 71 GPa. According to equation (3.31), therefore, for any given change in the value of the NOBP, the associated change in the porosity will be about 3.5 times larger in a shale than in a carbonate.

CALCULATION OF THE FRACTURE PRESSURE

The fracture pressure is the value of the pressure beyond which the structure of a rock begins to break down and is a quantity of both theoretical and practical interests. From a practical point of view, the pressure exerted by a column of drilling mud in a well should be less than the fracture pressure of the rocks, otherwise the rock may splinter and the drilling mud leak from the column. Mud lost in this way can potentially derail the drilling operation itself.

Here, we will first consider the original formula due to Hubbert and Willis (1957) for calculating the fracture pressure. This formula is based on the hypothesis that, in geological areas where normal faulting is prevalent, the effective horizontal stress must be less than the effective vertical stress (or NOBP).

The effective horizontal stress σ_H is defined in a similar way to the NOBP:

$$\sigma_H = S_H - PP,\tag{3.32}$$

where S_H is the horizontal stress and PP is the pore pressure.

The horizontal stress is more difficult to visualize than the vertical stress (*i.e.* the overburden pressure). Hubbert and Willis identify the horizontal stress with the fracture pressure FP, so that equation (3.32) can be rewritten as:

$$\sigma_H = FP - PP.\tag{3.33}$$

On the other hand, the effective vertical stress, which is equal to the NOBP, is given by the standard formula:

$$\sigma_V = OP - PP,\tag{3.34}$$

where, as before, OP is the overburden pressure.

As the fracture pressure can be measured in Leak-off tests, and the overburden and the pore pressure are also measureable quantities, Hubbert and Willis noted the empirical fact that for wells in the Gulf of Mexico:

$$\frac{FP - PP}{OP - PP} = \frac{\sigma_H}{\sigma_V} \cong \frac{1}{3}. \tag{3.35}$$

It should be emphasized that Hubbert and Willis did not use the theory of elasticity to generate equation (3.35).

THE EATON FORMULA (1969, 1997) FOR CALCULATING THE FRACTURE PRESSURE

If an elastic body is compressed inside a rigid steel cylinder, so that the body is prevented from expanding laterally, the relationship between the effective horizontal and effective vertical stresses is given by:

$$\frac{\sigma_H}{\sigma_V} = \frac{v}{1-v}, \tag{3.36}$$

where v is the Poisson ratio.

For elastic materials, the Poisson ratio ranges between 0 and 0.5, with 0.25 as a rough average. If v is assumed to be 0.25, the quotient on the right-hand side of equation (3.36) is equal to 1/3. Eaton inferred from this that the factor 1/3 which appears on the right-hand side of the Hubbert-Willis equation (3.35) should be replaced by $v/(1-v)$. Equation (3.35) then becomes:

$$\frac{FP - PP}{OP - PP} = \frac{v}{1-v}. \tag{3.37}$$

On the basis of fracture pressure data from Leak-off tests, Eaton established correlations between the Poisson ratio and the depth, and as more data became available, the correlations were improved. The most recent empirical correlations are as follows (with Z the depth in feet below the sea floor):

a.   For depths between 0 and 4999 feet below the sea floor:

$$v = -7.5E(-90)Z^2 + 8.0214286E(-05)Z + 0.2000714857. \tag{3.38a}$$

b.   For depths equal to or greater than 5000 feet below sea floor:

$$v = -1.7728E(-10)Z^2 + 9.4748424E(-06)Z + 0.3724340861. \tag{3.38b}$$

Figure 3.13 plots the correlations between the Poisson ratio and the depth given by equations (3.38a) and (3.38b). These correlations were established using equation (3.37) for those points where the fracture pressure was known (from Leak-off tests), as well as data on the overburden and the pore pressures.

However, there are some problems with Eaton's interpretation of the Hubbert-Willis formula (equation 3.35) in terms of the Poisson ratio. As can be seen in Figure 3.13, the Poisson ratio should theoretically increase with depth. However, in practice exactly

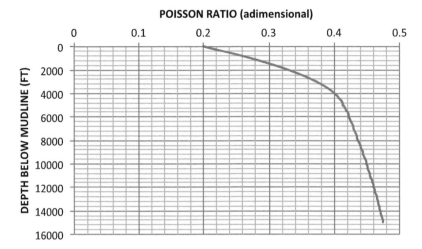

Figure 3.13 Eaton's correlation between the Poisson ratio (on the x axis) and the depth below the mudline in feet (on the y axis). Note the systematic increase of the Poisson ratio as the depth increases.

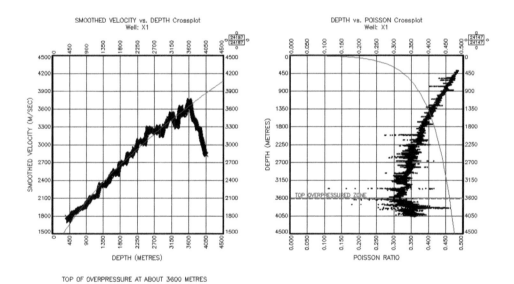

Figure 3.14 Left side: Velocity vs. depth plot, with the NCTL shown. A possible zone of very high pressure starts at about 3600 m. Right side: Depth vs. Poisson ratio, for the same well, showing the clear decrease of the Poisson ratio with depth. The faint smooth curve represents Eaton's correlations.

the opposite happens: namely, the Poisson ratio decreases with depth. The left-hand frame in Figure 3.14 plots the velocity (on the y axis) against the depth (on the x axis) plot taken from well data. An NCTL has been adjusted to fit the data and suggests that most of the well is at a normal, hydrostatic pressure. At about 3600 m there is a sharp decline in the velocity, which possibly indicates the start of a high-pressure zone. The right-hand frame in Figure 3.14 plots the depth (on the y axis, increasing downwards) against the Poisson ratio (on the x axis), for the same well. The Poisson ratio has been calculated from the DTCO (slowness of the compressional wave), the DTSM (slowness of the shear wave) and the Density logs. These then measure the actual values of the Poisson ratio, which shows a decrease with depth (in both the normally pressured and the overpressured sections). The faint line represents Eaton's correlations.

It appears that Eaton equation (3.37) does not have a consistent physical interpretation. The decrease in the Poisson ratio with depth, calculated on the basis of the DTCO, DTSM and Density logs, has been observed in countless wells. Yet, because equations (3.37), (3.38a) and (3.38b) are also based on field data involving the fracture pressure, they are well suited to applications that avoid any direct reference to the Poisson ratio. In the experience of the authors, the Eaton equations are excellent predictors of fracture pressure and their use is highly recommended.

Solving (3.37) for the fracture pressure FP gives:

$$FP = PP + (OP - PP)\frac{v}{1-v}, \tag{3.39}$$

where v is calculated at each depth Z, using equations (3.38a) and (3.38b). The parameter v should not be interpreted as the Poisson ratio, but rather as an undefined variable that fixes the right-hand side of equation (3.37). Whatever name or interpretation might be given to v, it remains an excellent predictor of the fracture pressure.

AN OIL EXPLORATION APPLICATION OF PORE PRESSURE

Sealing and non-sealing faults

The entrapment of oil against a fault is a fairly common occurrence. Many oilfields are formed by traps of this type. However, there are many situations in which potential oil traps yield disappointing results. If a possible trap is detected seismically, but the fault is not sealing, the consequences in terms of failed exploration costs can be dire. However, it is possible in certain circumstances to determine if a fault is sealing.

Figure 3.15 shows a seismic section, with colours corresponding to the interval velocities superposed on it. At about 2000 ms (and towards the right side of the figure), there is a sharp contrast in the velocities. There appears to be a significant fault, plunging to the right and separating the two zones. The obvious inference is that the two zones are at different pressures. If two points are at the same elevation but have different pressures, it is likely that the two points are not hydraulically connected. If, furthermore, there is a fault separating the points, there is good reason to believe that the fault is sealing. Hence, there could be an oil trap in this location.

The same contrast can be seen towards the left side of the figure, where a well was drilled and discovered gas. The presence of a fault between the two different zones is not as clear in this case.

This is just one example of the potential use of pore pressure measurements in exploration.

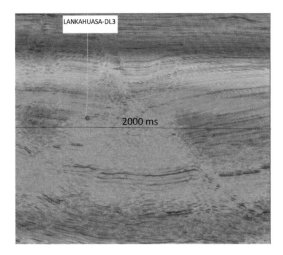

Figure 3.15 A seismic section which has been coloured with the interval velocities (the "hotter" colours imply higher velocities). Note the great contrast in velocities at about 2000 ms, with an apparent fault separating the two zones. The fault plunges to the right. The different velocities suggest two different fluid pressure regimes. Is this an indication that the fault is sealing and keeping the pressures apart?

REFERENCES

Eaton, B. (1969), Fracture Gradient Prediction and its Application in Oilfield Operation, JPT (October 1969) 1353; Trans. AIME, 246, in SPE Reprint Series No.49, Pore Pressure and Fracture Gradients.

Eaton, B. and Eaton, T. (1997), Fracture gradient prediction for the new generation, World Oil (October 1997) 93, in SPE Reprint Series No.49, Pore Pressure and Fracture Gradients.

Gluyas, J. and Cade, C. (1997), Prediction of Porosity in compacted sands, AAPG Memoir 69, Reservoir Quality prediction in Sandstones and carbibates.

Hottman, C.E. and Johnson, R.K. (1965), Estimation of formation pressures from log-derived shale properties; SPE Paper No.1110, in SPE Reprint Series No.49, Pore Pressure and Fracture Gradients.

Hubbert, M. and Willid, D. (1957), Mechanics of hydraulic fracturing, Trans. *AIME*, 210, 253.

Pettijohn, F., Potter P. and Siever, R. (1973), *Sand and Sandstone*, Springer Verlag, New York, Heidelberg, Berlin.

Zhang, J. (2013), Effective Stress, porosity, velocity and abnormal pore pressure prediction accounting for compaction disequilibrium and unloading, *Marine and Petroleum Geology*, 45, 2–11.

Chapter 4

Incompressibility of rocks and the Gassmann equation

INCOMPRESSIBILITY MODULI AND THE RELATIONSHIPS BETWEEN THEM

A non-porous elastic solid has just one type of incompressibility. However, rocks are porous solids and are characterized by a pore volume V_p as well as a bulk volume V_b. (As before, the porosity is defined to be $\phi = V_p/V_b$.) Furthermore, rocks are subject to both an external hydrostatic or overburden pressure, and an internal pore pressure. Following Zimmerman (1991), it is therefore possible to define at least four different incompressibilities, of which the first is:

$$\frac{1}{K_{bc}} = C_{bc} = -\frac{1}{V_b}\left(\frac{\partial V_b}{\partial P_c}\right)_{Pp} \tag{4.1}$$

Here, the incompressibility K_{bc} and its inverse, the compressibility C_{bc}, measure the fractional change in the bulk volume, as the confining pressure is varied but the pore pressure remains constant. The first letter in the subscript on the incompressibility or compressibility symbols can be either a "b", denoting the bulk volume, or a "p", denoting the pore volume. The second letter can be either a "c", if it is the pressure that varies, or a "p", if it is the pore pressure that varies. This allows three other possibilities, which are:

$$\frac{1}{K_{bp}} = C_{bp} = -\frac{1}{V_b}\left(\frac{\partial V_b}{\partial P_p}\right)_{Pc} \tag{4.2}$$

$$\frac{1}{K_{pc}} = C_{pc} = -\frac{1}{V_p}\left(\frac{\partial V_p}{\partial P_c}\right)_{Pp} \tag{4.3}$$

and

$$\frac{1}{K_{pp}} = C_{pp} = -\frac{1}{V_p}\left(\frac{\partial V_p}{\partial P_p}\right)_{Pc} \tag{4.4}$$

DOI: 10.1201/9781003261773-5

The signs in these equations have been chosen so that the incompressibility (or compressibility) is always positive. For instance, in equation (4.1), an increase in the confining pressure results in a reduction of the bulk volume, so the derivative without the minus sign would be negative. In equation (4.4), by contrast, an increase in the pore pressure generates an increase in the pore volume.

The incompressibility K_{bc} is also often written as "K_{dry}". It can be interpreted as the compressibility of a rock when its pore volume is devoid of any fluid. It is a critical parameter because it appears in the Gassmann equation. The incompressibility K_{pp} also plays an important role in reservoir engineering, where it is generally known as the pore compressibility.

In addition to the four incompressibilities defined above, there are two more that will be useful in what follows. These are:

a. The incompressibility K_o of the solid part of the rock (also known as the grain incompressibility)

$$\frac{1}{K_0} = C_r = -\frac{1}{V_b}\frac{dV_b}{dP_c} = -\frac{1}{V_b}\frac{dV_b}{dP_p} \tag{4.5}$$

whose inverse, C_r, is the compressibility of the solid part of the rock; and

b. The bulk or undrained bulk incompressibility

$$\frac{1}{K_b} = C_b = -\frac{1}{V_b}\left(\frac{\partial V_b}{\partial P_c}\right)_{MFLUID} \tag{4.6}$$

which measures the change of bulk volume as the confining pressure varies, while the mass of fluid occupying the pore volume remains constant.

In order to better understand the relationship between some of these incompressibility parameters, we refer the reader to Figure 4.1. In this figure, the x-axis indicates the pore pressure and the y-axis represents the confining or overburden pressure. The net overburden pressure, σ, is given by the difference between the confining pressure and the pore pressure. That is, $\sigma = P_c - P_p$, or equivalently $P_c = P_p + \sigma$. This formula has been plotted for several fixed values of the net overburden pressure, on the assumptions that the rock is perfectly elastic, and the pore volume V_p and the bulk volume V_b are state functions.

Let us investigate first the relationships between C_{bp}, C_{bc} and C_r.

Consider Point 1, where the pore pressure is 6 kgf/cm^2 and the confining pressure is 10 kgf/cm^2, on the iso-net overburden line of 4 kgf/cm^2. Suppose that the pore and confining pressures are changed so that the system moves to Point 2, and then to Point 3. What would be the change in the bulk volume?

In going from Point 1 to Point 2, the pore pressure remains constant and only the confining pressure changes, whereas in going from Point 2 to Point 3 only the pore pressure changes, while the confining pressure remains constant. So the net change in the bulk volume in moving Point 1 to Point 3 via Point 2 is:

$$dV_b = -V_b C_{bc} dP_c + V_b C_{bp} dP_p \tag{4.7}$$

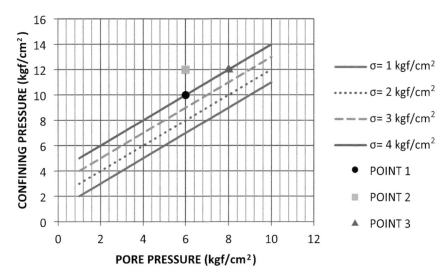

Figure 4.1 Lines of equal net overburden pressure, with the pore pressure along the x-axis and the confining pressure along the y-axis.

On the other hand, if the system moves from Point 1 to Point 3 directly, following the iso-net overburden pressure line, the change in bulk volume can be calculated from either of these two equations:

$$dV_b = -V_b C_r dP_c \qquad (4.8)$$

$$dV_b = -V_b C_r dP_p \qquad (4.9)$$

Note that from the definition (equation 4.5) of C_r, $dP_c = dP_p$. This condition is satisfied along any iso-net overburden pressure line, such as the direct trajectory from Point 1 to Point 3.

Because V_b should be a function only of the two pressures, the change in the bulk volume described by equation (4.7) should be the same as the change described by either equation (4.8) or equation (4.9). It follows that

$$-V_b C_r dP_c = -V_b C_{bc} dP_c + V_b C_{bp} dP_p \qquad (4.10)$$

and

$$-V_b C_r dP_p = -V_b C_{bc} dP_c + V_b C_{bp} dP_p \qquad (4.11)$$

After dividing equation (4.10) by equation (4.11) and solving for C_{bp}, we can conclude finally that:

$$C_{bp} = C_{bc} - C_r \qquad (4.12)$$

Following a similar line of reasoning, it can be shown that:

$$C_{pp} = C_{pc} - C_r \tag{4.13}.$$

Two further relationships between the compressibility (or incompressibility) parameters have been derived by Zimmerman (1991). These are:

$$C_{pc} = \frac{C_{bc} - C_r}{\phi} \tag{4.14}$$

and

$$C_{pp} = \frac{C_{bc} - C_r(1+\phi)}{\phi} \tag{4.15}$$

where ϕ is the porosity.

Equations (4.12)–(4.15) are important for an understanding of the derivation of the Gassmann equation. In theory, for the purposes of reservoir engineering, equation (4.15) allows the "pore compressibility" C_{pp} to be calculated from data taken from well logs. However, calculating the values of C_{bc} and C_r in this way implicitly assumes that the rock is elastic, whereas the depletion of a reservoir is not an elastic process. The values of C_{pp} taken from well logs will therefore differ from the values determined experimentally in the laboratory.

THE GASSMANN EQUATION

The Gassmann equation is without doubt one of the most important formulas in rock physics. The formula allows us to make what is in effect a "fluid substitution". If a sample of water-saturated rock is available whose velocities (longitudinal and shear) and density have been determined with well logs, the Gassmann equation tells us what the velocities would be if the rock were instead saturated by gas. And, of course, the substitution can be reversed. If we have a gas- or oil-saturated rock, with known values of the velocities, we can use the formula to estimate the velocities if the rock were water-saturated. These applications are important in AVO (Amplitude Variations with Offset) and inversion studies. It is therefore worthwhile to examine in some detail the origin of the equation and understand what its limitations are.

The Gassmann equation reads:

$$K_b = K_{dry} + \frac{\left(1 - \dfrac{K_{dry}}{K_o}\right)^2}{\dfrac{\phi}{K_f} + \dfrac{1-\phi}{K_o} - \dfrac{K_{dry}}{K_o^2}} \tag{4.16}$$

where K_b is the bulk incompressibility as defined in equation (4.6), $K_{dry} = K_{bc}$ as defined in equation (4.1), K_o is the incompressibility of the solid part of the rock, defined in equation (4.5), K_f is the fluid incompressibility and ϕ is again the porosity.

To derive Gassmann equation, we will follow an approach similar to Zimmerman's (1991).

We refer the reader again to Figure 4.1, where the system is assumed to move from Point 1 to Point 3, passing through Point 2. The changes in the bulk volume V_b and the pore volume V_p are given by:

$$\frac{dV_b}{V_b} = C_{bp}dP_p - C_{bc}dP_c \tag{4.17}$$

and

$$\frac{dV_p}{V_p} = C_{pp}dP_p - C_{pc}dP_c \tag{4.18}$$

If it is assumed that the whole pore space is occupied by a single fluid and that the pore pressure is constant throughout the pore volume, the fluid compressibility C_f is defined analogously to C_r by the equation:

$$\frac{dV_p}{V_p} = -C_f dP_p \tag{4.19}$$

From the definition (equation 4.6) of the bulk compressibility C_b, we have:

$$\frac{dV_b}{V_b} = -C_b dP_c \tag{4.20}$$

Also, from equations (4.18) and (4.19):

$$-C_f dP_p = C_{pp}dP_p - C_{pc}dP_c \tag{4.21}$$

while from equations (4.17) and (4.20):

$$-C_b dP_c = C_{bp}dP_p - C_{bc}dP_c \tag{4.22}$$

Rearranging both equations (4.21) and (4.22) gives

$$\left(C_{pp} + C_f\right)dP_p = C_{pc}dP_c \tag{4.23}$$

$$C_{bp}dP_p = \left(C_{bc} - C_b\right)dP_c \tag{4.24}$$

and dividing equation (4.23) by equation (4.24) so as to eliminate the differentials, then solving for C_b generates the final result:

$$C_b = C_{bc} - \frac{C_{bp}C_{pc}}{C_{pp} + C_f} \tag{4.25}$$

Equation (4.25) is the Gassmann equation. In view of equations (4.12), (4.14) and (4.15), C_{bp}, C_{pc} and C_{pp} can be replaced in equation (4.25) by C_{bc}, C_r and ϕ, and if the incompressibilities are used instead of the compressibilities and it is remembered that $K_{bc} = K_{dry}$, an equivalent version is:

$$\frac{1}{K_b} = \frac{1}{K_{dry}} - \frac{\left(\dfrac{1}{K_{dry}} - \dfrac{1}{K_o}\right)^2}{\dfrac{\phi}{K_f} + \dfrac{1}{K_{dry}} - \dfrac{(1+\phi)}{K_o}} \tag{4.26}$$

This is the Gassmann equation expressed in terms of incompressibilities (as conventionally used in rock physics). It can be rearranged to give formula (equation 4.16), which is the version that appears in most textbooks.

The only assumptions that underlie this equation are that the rock is fully saturated with a single fluid (although, as will be seen later, it is possible to account for the presence of more than one type of fluid) and that the pore pressure is constant in the volume of the rock under consideration. The second condition requires that the pores are interconnected, and hence that the rock is permeable to a certain degree. Provided that these conditions are satisfied, equation (4.16) can be used to calculate K_b from incompressibility measurements in the laboratory (which is to say, measurements taken while physically compressing the samples). However, some discrepancies appear when sonic measurements (the values of the primary and shear velocities) are used to calculate K_b. When a high frequency p-wave passes through a rock, the pore volume may not be at a uniform pressure and this is one of the reasons why there might be a discrepancy between theory and practice. However, for seismic frequencies in the order of 20–30 Hz, these discrepancies should be negligible. In sonic logging, the waves frequencies are usually much higher (perhaps ten or more times greater than the seismic frequencies), but it is still an acceptable industry practice to use sonic velocities in conjunction with the Gassmann equation.

THE RELATIONSHIP BETWEEN THE POROSITY AND THE NET OVERBURDEN PRESSURE

We will now demonstrate that, in the parameter space spanned by the pore pressure and the overburden pressure (as represented by Figure 4.1), the iso-porosity lines are parallel to the iso-net overburden pressure lines. This entails that the maximum change in porosity occurs if the system moves normally to the iso-net overburden pressure lines.

By definition, the net overburden pressure is given by:

$$\sigma = P_c - P_p \tag{4.27}$$

It follows therefore that

$$grad(\sigma) = \frac{\partial \sigma}{\partial P_p} I + \frac{\partial \sigma}{\partial P_c} J = -I + J \tag{4.28}$$

where I and J are unit vectors along the x- (pore pressure) and y- (confining pressure) axes, respectively.

Since $\phi = \dfrac{V_p}{V_b}$, the gradient of the porosity is given by:

$$grad(\phi) = \phi\left(C_{pp} - C_{bp}\right)I + \phi\left(C_{bc} - C_{pc}\right)J \qquad (4.29)$$

Taking the dot product of equations (4.28) and (4.29) then yields:

$$\left(C_{bp} - C_{pp}\right) + \left(C_{bc} - C_{pc}\right) = \sqrt{2}\sqrt{\left(C_{bp} - C_{pp}\right)^2 + \left(C_{bc} - C_{pc}\right)^2}\,\cos(\alpha) \qquad (4.30)$$

where α is the angle between the two gradient vectors.

From equations (4.12) and (4.13), it follows that:

$$\left(C_{bp} - C_{pp}\right) = \left(C_{bc} - C_{pc}\right) \qquad (4.31)$$

After substituting equation (4.31) into equation (4.30), it is evident that $\cos(\alpha) = 1$, which implies that $\alpha = 0$. In other words, both gradients are parallel, and as a consequence the iso-net overburden pressure lines are parallel to the iso-porosity lines. In the plane spanned by the pore pressure and the overburden pressure, the maximum change in the porosity therefore occurs if the system moves parallel to the gradient of the net overburden pressure.

What follows is an explicit proof that the porosity is controlled by the net effective overburden pressure $\sigma = P_c - P_p$.

Equation (4.29) is equivalent to:

$$d\phi = \phi\left[\left(C_{pp} - C_{bp}\right)dP_p + \left(C_{bc} - C_{pc}\right)dP_c\right] \qquad (4.32)$$

and if equation (4.31) is substituted into equation (4.32), it follows after some manipulation that

$$d\phi = \phi\left(C_{bc} - C_{pc}\right)\left(dP_c - dP_p\right) \qquad (4.33)$$

But $d\sigma = \left(dP_c - dP_p\right)$, so equation (4.33) demonstrates that the porosity is controlled by the net overburden pressure, which is what we set out to prove.

Furthermore, if equation (4.14) is substituted into equation (4.33), and it is remembered that $C_{bc} = C_{dry}$, then after replacing all the compressibilities with the corresponding incompressibilities equation (4.33) becomes:

$$d\phi = \left[\frac{K_{dry} - K_o\left(1 - \phi\right)}{K_o K_{dry}}\right]d\sigma \qquad (4.34)$$

Equation (4.34) is important because it expresses the dependence of changes in the porosity on changes in the net overburden pressure in terms of just the incompressibilities and the porosity, and despite the implicit assumption that the rock is elastic, it provides a theoretical background against which the empirical formulas of Bowers and Eaton for calculating the pore pressure from sonic data can be justified (in a non-elastic environment).

Furthermore, equation (4.34) provides a constraint on the value of K_{dry}, as can be seen as follows. In the case of a rock that is absolutely incompressible, $_s$, and therefore:

$$K_{dry} = K_o \left(1 - \phi\right).$$

For a given porosity, this formula gives the maximum possible of K_{dry}. However, rocks are not incompressible, and the derivative of the porosity with respect to the net overburden pressure must always be negative (as when the net overburden pressure increases, the porosity decreases). Hence, it must be true that

$$K_{dry} < K_o \left(1 - \phi\right).$$

SUMMARY

In this chapter, several incompressibility moduli have been defined, and used to derive the Gassmann equation, which has important applications in rock physics. Again using the incompressibility moduli, it has been demonstrated that the lines of iso-porosity in a state diagram spanned by P_p and P_c are parallel to the lines of iso-net overburden pressure. An equation has been derived which expresses any change in the porosity of a rock in terms of changes in the net overburden pressure. This equation places a constraint on the possible values of the parameter K_{dry} and in a sense justifies the empirical formulas that are conventionally used to calculate pore pressure from sonic or seismic data.

REFERENCE

Zimmerman, R. (1991), *Compressibility of Sandstones*, Elsevier, Amsterdam-Oxford-New York-Tokyo.

Chapter 5

Fluid substitution

THE FLUID SUBSTITUTION PROBLEM

From the point of view of seismic exploration, the central importance of rock physics lies in its capacity to generate models of the physical properties of rocks. Nowadays, both the P- and S-wave velocities are measured in nearly all exploration holes, and the Resistivity, Neutron, Density and Gamma logs are also recorded by default. If the lithology in a clastic stratigraphic column is sand and/or shale (which is to say, the petrophysical "shale" described in Chapter 1), it is a relatively straightforward matter to determine the elastic parameters at each point in the column. Furthermore, it is usually possible to classify the matrix as not just quartz, but as a mixture of quartz and another mineral (say calcite or dolomite).

A common theoretical problem, in a well where no hydrocarbons have been discovered, is to calculate the likely values of the two seismic velocities and the density if the sand was filled by a hydrocarbon rather than by water. This is a "fluid substitution" exercise, and to carry it out we need to first know the effective porosity and the lithological composition (the proportion of shale and matrix) of the column as well as the original hydrocarbon saturation. We also need to estimate the incompressibility of the prevailing mixture of solids (where the "shale" is considered to be a solid mineral, *cf.* Chapter 1). To do this, it is useful to introduce upper and lower bounds on the elastic parameters in the mixture. The standard bounds used are either the Reuss lower bound and the Voigt upper bound, or the Hashin-Shtrikman upper and lower bounds. A final necessary piece of information is the physical properties of water, oil and gas at the reservoir pressure and temperature. Armed with these, it should in principle be possible to use the Gassmann equation to calculate the bulk incompressibility of the rock if its fluid content were different. As will be discussed at the end of the chapter, there is a second, alternative method that can be used to estimate the bulk incompressibility. Note that a change in the fluid content of a rock will not change its shear *modulus*, but its shear *velocity* will change as a result of the change in density.

PHYSICAL PROPERTIES OF FLUIDS

The two properties of interest in this chapter are the incompressibility and density of the confined fluids, in a situation where a single fluid or two or more fluids partially

DOI: 10.1201/9781003261773-6

occupy the pore space in the rock. The effective fluid incompressibility K_f when two fluids (water and hydrocarbons) fill the pore space is given by:

$$\frac{1}{K_f} = \frac{S_w}{K_w} + \frac{1-S_w}{K_h} \tag{5.1}$$

where K_w and K_h are the incompressibilities of the water and the hydrocarbon fraction, and S_w is the water saturation. When $S_w = 1.0$, the effective fluid incompressibility is equal to K_w. For the purposes of quick calculations, K_w can be assumed to be 2.3 GPa (the incompressibility of pure water in standard conditions is 2.15 GPa). A water density of 1.03 g/cc would be a fair approximation. In reality, the incompressibility and density of water depend on its pressure, temperature, salinity and the amount of dissolved gas. Also, the shear modulus of a fluid is zero.

For quick calculations, the incompressibility of oil can be assumed to be about 1.0 GPa, and the density of oil in reservoir conditions as 0.8 g/cc. These are typical approximate values, as both the incompressibility and density of a hydrocarbon depend on its pressure, temperature, gravity and gas/oil ratio.

The incompressibility K_g and density ρ of a gas can both be quickly calculated from the ideal gas equation of state, which reads as follows:

$$\rho = \frac{M_w P}{RT} \tag{5.2}$$

where R is the universal gas constant (equal to 0.082 l atm/(mole K)), T is the temperature (in K), P is the pressure (in atm) and M_w is the molecular weight of the gas (in g).

By definition, the incompressibility of the gas (if it is ideal) is given by

$$\frac{1}{K_g} = -\frac{1}{V}\left(\frac{\partial V}{\partial P}\right)_{T=const}, \tag{5.3a}$$

but

$$\left(\frac{\partial V}{\partial P}\right) = -\frac{nRT}{VP^2} \tag{5.3b}$$

From equations (5.3a) and (5.3b), it follows that:

$$K_g = P \tag{5.3c}$$

That is to say, the incompressibility of an ideal gas is equal to its pressure.

Mavko et al. (1998) have provided a set of correlations which can be used to estimate the density and incompressibility of water, oil and gas under most pressure and temperature conditions. In the case of water, the correlations require a measurement of the salinity, while for gas a value of the gas gravity is needed. For oil, the necessary inputs are the associated gas gravity, the gas/oil ratio and the API gravity.

Calculating the incompressibility and density of single fluids is a rather laborious process (cf. Mavko et al. 1998), unless the analyst is in possession of software pre-programmed to perform the calculations. Considering all the uncertainties associated with the fluid substitution exercise, it is worthwhile asking if the supposedly "exact" numbers furnished by the correlations are really necessary, or can be replaced by the quick approximations described above. If necessary, the correlation calculations can be performed using EXCEL. In particular, in the case of gas it is convenient to calculate the density and incompressibility exactly, even if we know nothing about the expected composition. The gas gravity of methane (*i.e.* the ratio of the density of methane to the density of air at 1 atm and 60 F) is 0.56, and ordinarily natural gas is 80%– 90% methane. Consequently, the gravity of the gas mixture should not be very different from 0.56. The pressures and temperatures in rock formations of interest are usually well reported, which furnishes almost all the data needed to apply the correlations. It is more difficult to apply the correlations to water and oil if a lot of information is missing, such as the gravity, the gas/oil ratio or the associated gas specific gravity for the oil, and the salinity or amount of dissolved gas for the water.

At this stage, it is instructive to understand how even a small gas saturation can change the effective incompressibility of the fluid, as will be seen if the effective fluid incompressibility is calculated for the following set of data:

Pressure: 100 kgf/cm^2
Temperature: 30°C
Gas gravity: 0.65
Water salinity: 50,000 ppm (NaCl)
Water incompressibility from correlations: 2.56 GPa
Assumption of water incompressibility: 230 GPa
Gas incompressibility: 0.0164 GPa (from correlations)
Gas incompressibility: 0.0098 GPa (assuming an ideal gas)

Solving equation (5.1) for K_f gives:

$$K_f = \frac{K_w K_g}{K_g S_w + (1 - S_w) K_w} \tag{5.4}$$

Figure 5.1 plots K_f (the effective incompressibility of the two-phase fluid) as a function of the water saturation, using the two sets of parameters (those calculated from the correlations, and the approximations). Two features of the graph should be noted. First, there is little difference between the curve generated from the approximate parameters and the curve based on the parameters derived from the correlations. Second, the presence of even a very small saturation of gas (of the order of 1%) causes a drastic change in the effective fluid compressibility. However, we still need to calculate the effect of changing the fluid composition on the two seismic velocities and the overall density of the rock. This involves a fluid substitution exercise.

A SIMPLE FLUID SUBSTITUTION EXERCISE

The exercise will be "simple" because it involves a "clean" sand, for which the elastic parameters of the matrix (quartz) are well known. In a more realistic case, the rock

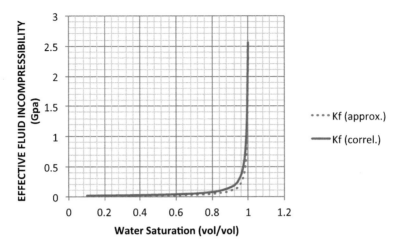

Figure 5.1 Effective fluid incompressibility as a function of water saturation. In this case, there is little difference between the two curves, one constructed using the approximate parameters (the default procedure when knowledge of the properties of the fluid is scant) and the other using the parameters generated by the correlations, which should represent the "true" values. Note that even a gas saturation of 1% produces a drastic drop in the effective fluid incompressibility.

would contain other minerals aside from quartz, and the parameters of the composite solid would need to be calculated. We will address this second problem later.

Assume therefore that the sand is clean and 100% water-saturated, with the following characteristics:

Porosity: 0.25
V_p: 3898 m/s
V_s: 2494 m/s
Bulk density: 2.245 g/cc
Quartz incompressibility: 37 GPa
Quartz shear modulus: 44 GPa
Water incompressibility: 2.56 GPa
Water density: 1.03 g/cc

We will use the Gassmann equation (described in Chapter 4) to determine the values of the two velocities and the density if the same rock had a gas saturation of 0.70 and a water saturation S_w of 0.30. The relevant parameters for the gas are:

Gas incompressibility: 0.016 GPa
Gas density: 0.09 g/cc

All the properties of water and gas listed above are for standard reservoir conditions.

The fluid substitution calculation involves the following steps:

1. Calculate the bulk modulus (K_b) and the shear modulus (μ) of the wet rock

From the equations

$$V_p = \sqrt{\frac{K_b + \frac{4}{3}\mu}{\rho_b}} \qquad (5.5a)$$

$$V_S = \sqrt{\frac{\mu}{\rho_b}} \qquad (5.5b)$$

we can solve for K_b, giving:

$$K_b = \rho_b \left(V_P^2 - \frac{4}{3}V_S^2 \right) = 15.49\,GPa$$

Note that if we express the velocities in km/s and the density in g/cc, then K_b is by default in GPa.

The shear modulus is therefore:

$$\mu = \rho_b V_S^2 = 13.96\,GPa$$

2. Calculate the dry modulus (K_{dry})

From Chapter 4, the Gassmann equation can be written as:

$$K_b = K_{dry} + \frac{\left(1 - \dfrac{K_{dry}}{K_0}\right)^2}{\dfrac{\phi}{K_{fl}} + \dfrac{1-\phi}{K_0} - \dfrac{K_{dry}}{K_0^2}}$$

where K_0 is the incompressibility of the solid part of the rock (for pure quartz, $K_0 = 37\,GPa$), K_f is the fluid incompressibility (at this stage, the fluid is water with an estimated incompressibility of 2.56 GPa) and K_{dry} (defined in Chapter 4) is the inverse of the compressibility C_{bc} and can be regarded as the bulk modulus when the rock is "dry" (i.e. its pores do not contain any fluid at all). According to the Gassmann equation, $K_{dry} = K_b$ when $K_f = 0$, and more generally:

$$K_{dry} = \frac{K_0^2 K_f - K_b K_0 \left[K_0 \phi + (1-\phi)K_f \right]}{2 K_0 K_f - K_b K_f - K_0^2 \phi - (1-\phi)K_0 K_f}$$

It follows from this formula that K_{dry} is equal to 10.99 GPa.

Incidentally, this is a feasible value of K_{dry}, as it should be recalled from Chapter 4 that $K_{DRY} \leq K_o(1-\phi)$. The maximum possible value of K_{dry} is therefore 27.75 GPa, which is well above the actual value of 10.99 GPa.

3. Calculate the effective fluid compressibility and the density under the new fluid saturation conditions

The water saturation is 0.30. In order to calculate the effective fluid incompressibility, we apply equation (5.1) with $K_w = 2.56$ and $K_g = 0.016$ GPa. This results in a value of K_f equal to 0.023 GPa.

The density of the new fluid is calculated using the formula:

$$\rho_f = \rho_w S_w + (1 - S_w)\rho$$

where the water density ρ_w is 1.03 g/cc, and the gas density ρ_g is 0.09 g/cc. It follows that the effective fluid density ρ_f is 0.372 g/cc.

4. Given the new value of K_f, calculate the new value of the bulk incompressibility K_b of the rock, assuming 70% gas in the pore space

From the Gassmann equation, the new value of K_b is 11.035 GPa.

5. Calculate the new seismic velocities and the new bulk density

To calculate the new bulk density, we proceed as follows.

The bulk density for the original fluid composition is given by

$$\rho_{bold} = \rho_{ma}(1-\phi) + \rho_{fold}\phi, \tag{5.6a}$$

where ρ_{fold} is the density of the original fluid.

The bulk density with the final (new) fluid composition is:

$$\rho_{bnew} = \rho_{ma}(1-\phi) + \rho_{fnew}\phi, \tag{5.6b}$$

where ρ_{fnew} is the density of the final fluid.

Combining equation (5.8a) and equation (5.6b) gives:

$$\rho_{bnew} = \rho_{bold} - \phi(\rho_{fold} - \rho_{fnew}), \tag{5.6c}$$

where ρ_{bnew} is the new bulk density (the result of the fluid substitution), ρ_{bold} is the original bulk density (equal to 2.245 g/cc in this case), ρ_{fold} is the density of the original fluid (which was water, with an estimated density of 1.03 g/cc) and ρ_{fnew} is the new fluid density, calculated in step 3 and equal to 0.37 g/cc.

Table 5.1 The results of the fluid substitution exercise, showing the original and final
values of the velocities and the density.

	Units	Original	Fluid Subst.	Remarks Remarks
Fluid		$S_w = 1.0$	$S_w = 0.3$	$S_{gas} = 0.7$
V_p	m/s	3898	3775.44	Compr. Vel.
V_s	m/s	2494	2500 63	Shear Vel.
ρ_b	G/CC	2245	2.08	
Z_p	(m/s)(g/cc)	8751	7852.91	Impedance
V_p/V_s	adimensional	1.562	1.457	

Substituting these values into equation (5.6c) gives a final bulk density ρ_{bnew} of
2.08 g/cc.

Because the shear modulus does not change when the fluid content of the rock
changes, the compressional and the shear velocities, V_p and V_s, can be calculated from
equations (5.5a) and (5.5b). This gives:

$$V_p = 3775.44 \, \text{m/s and } V_s = 2590.63 \, \text{m/s}.$$

Table 5.1 summarizes the results of this exercise. Note that the quotient V_p/V_s and the
p-impedance (which is defined to be the compressional velocity times the bulk density)
are quite different for water sands and gas-bearing sands. It should therefore be possible to distinguish between water sand and gas sand from the values of V_p/V_s and/or the
p-impedance for the shale in an inversion.

REUSS LOWER BOUND AND VOIGT UPPER BOUND AND
HASHIN-SHTRIKMAN UPPER AND LOWER BOUNDS

The effective elastic constants (*i.e.* the mean values of these parameters for the rock
as a whole, taking into account all liquid and solid content) will depend on the corresponding values for the individual components of the rock. However, the effective
elastic constants depend on other factors as well, such as the rock texture, whether the
rock is laminated, whether it is well sorted, and so on. For this reason, there is always
some uncertainty in the values of the effective elastic constants of any rock.

It is clear that the uncertainty in the value of a variable can be reduced if some
information is available about the possible range of the variable. The Reuss lower
bound and the Voigt upper bound are the lowest and highest possible values a given
parameter may take. The Hashin-Shtrikman bounds are conceptually the same, but in
many cases are more precise than the Reuss-Voigt bounds. Quite frequently, the lower
Hashin-Shtrikman bound is greater than the Reuss bound, and the upper Hashin-
Shtrikman bound is smaller than the Voigt bound.

If M denotes the value of any elastic constant, and x_1 and $x_2 = 1 - x_1$ are the volume
fractions of component 1 and component 2 in a two-component mixture, the Voigt and
Reuss bounds are given by the formulas:

$$M_{VOIGT} = M_1 x_1 + M_2(1 - x_1) \tag{5.7}$$

$$\frac{1}{M_{REUSS}} = \frac{x_1}{M_1} + \frac{(1-x_1)}{M_2} \tag{5.8}$$

Mavko et al. (1998) have pointed out that when the elastic parameters of the components of the rock are very different, these bounds are not precise enough to usefully describe the effective parameters of the rock.

Consider for example a case where M is the incompressibility and the rock is clean sand composed of pure quartz with 100% water saturation. The incompressibility of quartz is 37 GPa, whereas the incompressibility of water is about 2.3 GPa. One possible estimate of the effective incompressibility of the rock, or bulk incompressibility, is an average of the upper and lower bounds, which is also known as the Hill average (Mavko et al., 1998):

$$K_b = \frac{K_{bREUSS} + K_{bVOIGT}}{2} \tag{5.9}$$

where K_{bREUSS} and K_{bVOIGT} are given, respectively, by:

$$K_{bREUSS} = \frac{K_w K_{qtz}}{K_w(1-\phi) + \phi K_{qtz}} \tag{5.10}$$

$$K_{bVOIGT} = K_{qtz}(1-\phi) + K_w \phi \tag{5.11}$$

Here, K_{qtz} is the incompressibility of the quartz, K_w is the incompressibility of the water and ϕ is the porosity.

Figure 5.2 plots the upper and lower bounds and the estimated effective bulk incompressibility (the average of the two bounds) against the porosity. Because the

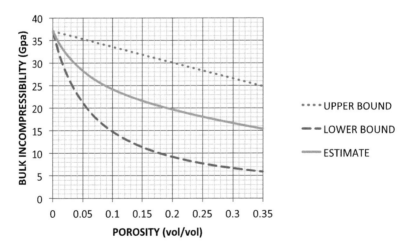

Figure 5.2 When the parameters of the individual constituents are very different (the incompressibility of quartz is 37 GPa, and the incompressibility of water is 2.3 GPa) the predictive power of the Voigt and Reuss bounds is poor, because the two bounds are widely separated.

difference between the incompressibility of quartz (37 Gpa) and the incompressibility of water (2.3 GPa) is substantial, the estimated value of the effective bulk incompressibility is quite uncertain. Any vertical line joining the two bounds for a given porosity shows the range of possible values of the effective bulk modulus.

It is evident from this example that it is not the best option to use the Reuss and Voigt bounds to predict the bulk modulus of a rock if it contains a mineral of high incompressibility together with water.

Let us now consider a slightly different problem. A rock contains calcite, dolomite and a pore space filled with water, and we would like to predict the incompressibility of the mixture of calcite and dolomite. The rock can be regarded as a binary system, comprising SOLID (calcite + dolomite) and FLUID (water). The incompressibility of calcite is 70.76 GPa, while the incompressibility of dolomite is about 80.23 GPa. The relative proximity of the two incompressibilities suggests that the average of the Reuss and Voigt bounds will give a good estimate of the incompressibility of the solid part of the rock (which would be the value of K_0 appearing the Gassmann equation). Figure 5.3 confirms this expectation. In this figure, the abscissa is the volume fraction of calcite in the solid part of the rock, which is defined by:

$$x = \frac{V_{CAL}}{1-\phi} \tag{5.12}$$

where V_{CAL} is the volume fraction of calcite in the rock as a whole and ϕ is again the porosity.

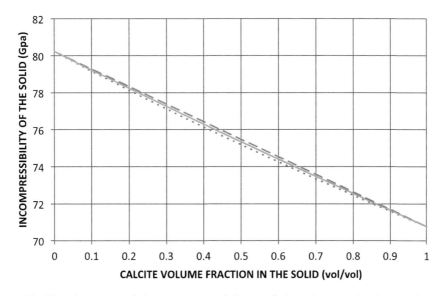

Figure 5.3 The closeness of the incompressibilities of the calcite and dolomite fractions leads to a small difference between the upper and lower bounds for the calcite-dolomite mixture, and so to a good prediction of the incompressibility of the solid part of the rock.

For binary compounds, the Hashin-Shtrikman bounds for the incompressibility and shear modulus are:

$$K^{HS\pm} = K_1 + \frac{f_2}{(K_2 - K_1)^{-1} + f_1\left(K_1 \frac{4}{3}\mu_1\right)^{-1}} \tag{5.13a}$$

$$\mu^{HS\pm} = \mu_1 + \frac{f_2}{(\mu_2 - \mu_1)^{-1} + \dfrac{2f_1(K_1 + 2\mu_1)}{5\mu_1\left(K_1 + \dfrac{4}{3}\mu_1\right)}} \tag{5.13b}$$

respectively, where f_k is the volume fraction of component k in the mixture. The upper (+) and lower (–) bounds correspond to the two choices available for identifying component 1 and component 2. In general, the upper bound results when component 1 is the stiffer of the two materials and the lower bound when component 2 is the stiffer material.

As an example, let us calculate the upper and lower bounds for the bulk modulus or effective bulk incompressibility of clean sand, composed of quartz with a porosity of 0.2% and 100% water saturated. If the quartz is chosen to be component 1 and the water to be component 2, then equation (5.14a) gives:

$$K_b^+ = K_{qtz} + \frac{\phi}{\left(K_w - K_{qtz}\right)^{-1} + (1-\phi)\left(K_{qtz} + \dfrac{4}{3}\mu_{qtz}\right)^{-1}} \tag{5.14}$$

On the other hand, if the water is chosen to be component 1 and the quartz to be component 2, then:

$$K_b^- = K_w + \frac{(1-\phi)}{\left(K_{qtz} - K_w\right)^{-1} + \phi K_w^{-1}} \tag{5.15}$$

(remembering that the shear modulus of a fluid is zero).

From equation (5.14), K_b^+ is 27.22 GPa, and from equation (5.15) K_b^- is 9.209 GPa. This entails that the true value of K_b should lie between 9.2099 and 27.22 GPa. Although the uncertainty in the value of the true bulk modulus is still substantial, the range of possible values is now marginally narrower than the range corresponding to the Reuss-Voigt bounds (which were 9.209 and 30.06 Gpa) (see Figure 5.4). Incidentally, if we compare equations (5.10) and (5.15), we see that at least for binary mixtures, if one of the components is a fluid, the Reuss bound and the Hashin-Shtrikman lower bound are the same.

MARION'S HYPOTHESIS

On the assumption that the true value M of any elastic parameter must lie between the Reuss and Voigt bounds (or alternatively, the Hashin-Shtrikman bounds, which are more precise), this true value can be represented in the form:

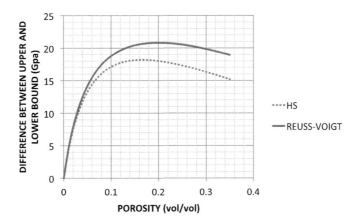

Figure 5.4 Difference between the upper and lower bounds for the bulk incompressibility of the quartz-water system, calculated using the Hashin-Shtrikman (lower
curve) and the Reuss-Voigt formulas (upper curve). For all possible values of
the porosity, the difference between the upper and lower bounds – and therefore the uncertainty in the true value of K_b – is smaller if the Hashin-Shtrikman
formula is used.

$$M = wM^- + (1-w)M^+ \qquad (5.16)$$

where M^- is the lower bound, M^+ is the upper bound and w is a weight lying in the
range [0, 1]. This formula ensures that the true value of the elastic constant lies between
the upper and lower limits. If $w = 0$, the true value of the constant is equal to the upper
bound, while if $w = 1$, the true value is equal to the lower bound.

Marion (in Mavko et al., 1998) has proposed the hypothesis that the value of the
weight w depends on the physical characteristics of the rock, such as its mineralogical
composition and texture, but is independent of the type of fluid filling the pore space.
Marion's hypothesis is useful because it offers an alternative to the Gassmann equation in the fluid substitution problem.

What follows is an experiment designed to check the accuracy of Marion's hypothesis that the weights are independent of the fluid composition in the pore space.
The experiment is not strictly speaking an unbiased test, because it assumes that the
Gassmann formula is valid and that K_{dry} can be calculated from a standard but empirical correlation with the porosity.

Assume that M in equation (5.16) is the bulk modulus K_b, that the rock is sand
whose sole mineral content is quartz and that the pore space is 100% water saturated.
Solving equation (5.16) for w gives:

$$w = \frac{K_{(w)}^+ - K_{b(w)}}{K_{(w)}^+ - K_{(w)}^-} \qquad (5.17)$$

In equation (5.17), $K_{(w)}^+$ is the upper bound for K_b, for a sand that is 100% water-
saturated, $K_{(w)}^-$ is the corresponding lower bound and $K_{b(w)}$ is the actual bulk modulus of the rock. As before, w represents the weight of the lower bound.

The analogous equation for the same rock, but with oil in the pore space instead of 100% water saturation, reads:

$$w = \frac{K_{(oil)}^+ - K_{b(oil)}}{K_{(oil)}^+ - K_{(oil)}^-} \tag{5.18}$$

According to Marion's hypothesis, the weight does not depend on the fluid, so the expressions on the right of equations (5.17) and (5.18) should be equal. If $K_{b(w)}$ is known, we can therefore calculate $K_{b(oil)}$.

To determine if the hypothesis actually works in practice, we will use equation (5.17) to calculate w, for the case of 100% water saturation, over a certain range of porosities, and equation (5.18) to calculate w, in the case where there is oil in the pore space, over the same range. The Gassmann equation will be used to calculate the two bulk moduli $K_{b(w)}$ and $K_{b(oil)}$, and the value of K_{dry} – which is needed for the Gassmann equation – will be calculated using function correlating K_{dry} with the porosity proposed by Murphy et al. for clean sands (which will be discussed in more detail in Chapter 6).

The values of w calculated from equations (5.17) and (5.18) in this way are plotted against each other in Figure 5.5. If the points had fallen on a straight line with slope 1 passing through the origin, Marion's hypothesis would have been confirmed perfectly in this case. Figure 5.5 shows the calculated data points, as well as the ideal straight line $w(\text{water}) = w(\text{oil})$. A line of best fit has been superimposed on the calculated data points, and this predicts that $w(\text{oil}) = -0.073$ when $w(\text{water}) = 0$ and that $w(\text{oil}) = 1.075$ when $w(\text{water}) = 1$. Both numbers are reasonably close to the expected values of 0 and 1.

The outcome of this particular experiment is that the Marion's hypothesis seems to work reasonably well, and indeed any deviation from the ideal straight line could in

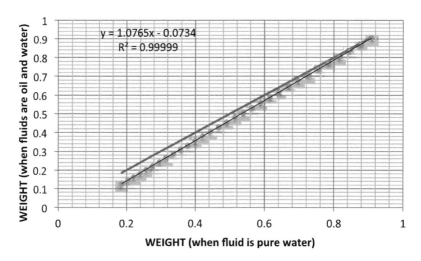

Figure 5.5 Testing Marion's hypothesis. The weights calculated for water-saturated sand are on the x-axis, and the weights calculated for mixed oil-water saturation are on the y-axis. If Marion's hypothesis were satisfied exactly, the points would lie on the upper line.

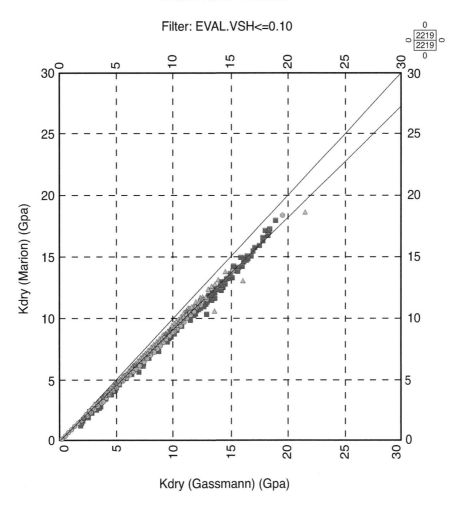

Kdry (Marion) vs. Kdry (Gassmann) Crossplot
DATA FROM 4 WELLS

Filter: EVAL.VSH<=0.10

Wells: ■ WELL No.1   ● WELL No.2   △ WELL No.3   ★ WELL No.4

CC: 0.997876
Kdry (Marion) = 0.906101*Kdry (Gassman)

Figure 5.6 Comparison of the values of K_{dry} calculated using the Gassmann equation and Marion's hypothesis, from field data for clean sands encountered in four wells. The two sets of values are very similar, which suggests that Marion's hypothesis is robust. The upper line is the 1:1, where all the points should fall if there were a perfect agreement between Marion and Gassmann. The lower line is the least-squares fit of the data.

theory be attributed to uncertainties in the Gassmann equation or the Murphy et al. correlation, which are just empirical approximations. In the absence of a proof from first principles, it is difficult to be certain whether Marion's hypothesis is an exact general law. However, the authors feel that it is a reliable empirical rule, and for this reason, it will be assumed to be true in all later sections.

If Marion's hypothesis is applied to the limiting case where there is no fluid in the pore space of the rock (*i.e.* when $K_f = 0$), the value of K_{dry} can be calculated independently of the Gassmann equation, as follows:

$$
K_{DRY} = \frac{K_b - K_{INF}}{K_{SUP} - K_{INF}} \left\{ K_0 + \frac{\phi}{(1-\phi)\left[K_0 + \frac{4}{3}\mu_0\right]^{-1} - \frac{1}{K_0}} \right\}
\tag{5.19}
$$

Here, K_{SUP} is the Hashin-Shtrikman upper bound for the bulk modulus K_b, K_{INF} is the H-S lower bound for K_b, μ_o is the shear modulus of the solid part of the rock and K_0 is again the incompressibility of the solid part of the rock.

If K_b is known, the value of K_{dry} can be calculated from equation (5.19), which therefore constitutes an alternative to the Gassmann equation.

The fundamental difference between equation (5.19) and the Gassmann equation is that equation (5.19) involves the shear modulus μ_0 of the solid part of the rock, which is absent from the Gassmann equation.

Figure 5.6 compares the values of K_{dry} calculated using the Gassmann equation and equation (5.19), based on field data from clean sand intervals (with shale volume fractions less than 0.10) in four wells, which had very complete sets of logs. The upper line is the theoretical straight line that would result if both methods gave identical numbers. The lower line is the line of best fit to the actual paired values. The proximity of the two lines suggests that Marion's hypothesis and the Gassmann equation give similar results.

In principle, the Gassmann equation and equation (5.19) can be combined to eliminate K_b. This would result in a single equation for K_{dry} as a function of the porosity, which could then be compared with the various empirical equations that are commonly used to calculate K_{dry} from porosity data. However, this topic is beyond the scope of the material considered here.

REFERENCE

Mavko, G., Mukerji, T. and Dvorkin, J. (1998), *The Rock Physics Handbook*, Cambridge University Press, Cambridge-New York-Melbourne.

Chapter 6

Forward modelling and empirical equations

FORWARD MODELLING AND EMPIRICAL EQUATIONS

If an exploration hole is to be drilled in an area where the stratigraphic column is expected to be a sand/shale sequence, it may be of interest to predict the physical properties of the sands that are the targets of the well (assuming there are no other wells in the vicinity of the exploration hole). At each depth, it is possible to make a rough estimate of the porosity of the sand. To assess the physical properties of the sand, it must be assumed that it is relatively clean, with no more than a small amount of shale. As a first step, also, it can be assumed that the sand is water-saturated. However, even given these ideal conditions – the lithology is a relatively clean sand, and the porosity and the water saturation are known – it is not possible to predict exactly the physical parameters we are most interested in, which are the compressional velocity V_p, the shear velocity V_s, and the bulk and shear moduli. In any forward modelling exercise, the only quantity that is known directly is the density of the sand. To estimate some of the other parameters, we need to resort to empirical formulas and correlations, whose validity might be limited to the data sets from which they were constructed. That is to say, there are no truly "universal" correlations, applicable in all circumstances. Eventually, one might use the Hertz-Mindlin model for granular media (Mavko et al., 1998), which allows the calculation of the elastic moduli of rocks consisting of uncemented spherical grains of very uniform radii. The scope of the application of the Hertz-Mindlin model is limited (rocks are generally somewhat cemented, the grains are not necessarily spherical and their radii is far from uniform) and will not be further discussed in this work.

In general, an exercise in forward modelling is first carried out with water sands. At a later stage, a fluid substitution exercise is performed to determine the physical properties of the sand if it is gas- or oil-saturated. This fluid substitution exercise will require calculation of the bulk and shear moduli (as the latter is used as an input when calculating the bulk modulus).

THE WYLLIE EQUATION

We will consider first a method for estimating V_p for a clean sand with a known porosity. To do this, we will use an empirical equation due to Wyllie et al. (1958). This equation has been extensively used in well log analysis, and its limitations have been described by Mavko et al. (1998). The equation is:

DOI: 10.1201/9781003261773-7

$$\Delta T = \Delta T_{MA}(1-\phi) + \Delta T_F \phi \qquad (6.1)$$

where the symbols ΔT refer to the transit time or slowness in the rock, which is the inverse of the compressional velocity. The subscripts "MA" and "F" refer to the matrix and the fluid, respectively. In terms of the corresponding velocities, the equation can be rewritten as:

$$\frac{1}{V_p} = \frac{(1-\phi)}{V_{MA}} + \frac{\phi}{V_F} \qquad (6.2)$$

For practical applications involving sands, ΔT_{MA} is generally taken to be 55 µs/ft, and ΔT_F to be 189 µs/ft.

Figure 6.1 shows a crossplot of porosity on the x-axis against the transit time or slowness on the y-axis for data taken from an interval about 1000 m thick. All the points on the graph have a shale content of less than 0.10, so effectively refer to clean sands. The porosity values have been calculated from the Neutron and Density logs, while the slowness values have been taken from the DTCO log. Three lines are visible in this plot: a thin straight line which represents the least-squares fit to the data, a thick straight line which is the standard Wyllie line (assuming that ΔT_{MA} = 55 µs/ft and ΔT_F = 189 µs/ft), and a thick curved line corresponding to the Raymer-Hunt-Gardner equation, which will not be discussed further in this book.

Note that the least-squares fit to the data has a relatively high correlation coefficient ($r^2 = 0.64$), which means that a linear relation such as equation (6.1) can adequately describe the relationship between the porosity and the transit time. However, the actual transit time parameters obtained from the regression line (ΔT_{MA} = 58.73 µs/ft and ΔT_F = 186.7 µs/ft) are quite different from the standard ones. It appears that, for this set of well data, the standard predictions for the velocity are biased away from the actual trend line.

Let us now make a rough estimate of the uncertainty associated with the value of the compressional velocity, calculated using the Wyllie equation. The standard formula assumes the values ΔT_{MA} = 55 µs/ft and ΔT_F = 189 µs/ft. Suppose that these parameters have no fixed values, but instead are random variables. For instance, we could assume that ΔT_{MA} is normally distributed, with a mean of 55 µs/ft and a variance chosen so that there is a 0.045 probability that ΔT_{MA} lies outside the range of 50–60 µs/ft (meaning that the odds that the value of ΔT_{MA} will stay inside that range is 95.5%). ΔT_F could also be treated as a normally distributed random variable, with a mean of 189 µs/ft and a variance chosen so that that the probability that ΔT_F lies outside the interval of 174–204 µs/ft is 0.045. According to the Wyllie equation (6.1), ΔT is the sum of two random variables and so is also a normally-distributed random variable.

The arbitrary assumption that the two parameters ΔT_{MA} and ΔT_F lie inside certain fixed ranges with 95.5% probability corresponds to a specification of their 4σ ranges. So the standard deviation in the probability distribution for ΔT_{MA} is σ_{MA} = 10/4 = 2.5 µms/ft, and for ΔT_F is σ_F = 30/4 = 7.5 µms/ft.

The corresponding probability distribution for ΔT itself has a porosity-dependent mean μ and variance σ^2 given by:

$$\mu = (1-\phi)\mu_1 + \phi\mu_2 = 55(1-\phi) + 189\phi \qquad (6.3a)$$

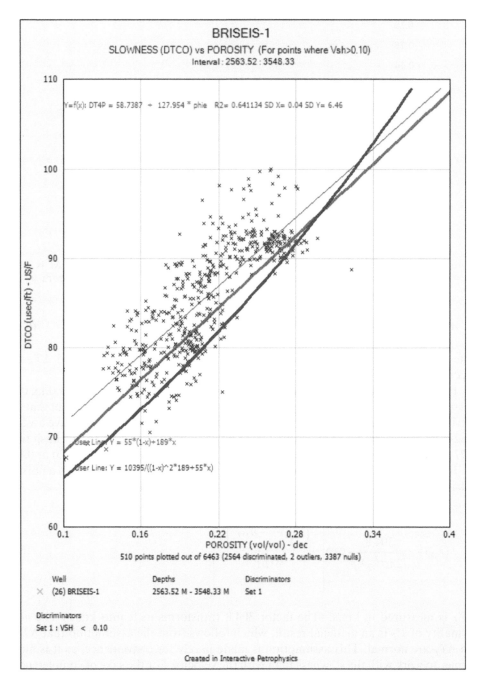

Figure 6.1 A crossplot of transit time vs. porosity for an interval about 1000 m thick, for points with $V_{sh} < 0.10$. The least-squares fit to the data (the thin line), the predicted correlation from the Wyllie equation (the thick line) and prediction of the Raymer et al. equation (the thick CURVED line) are all shown in the diagram.

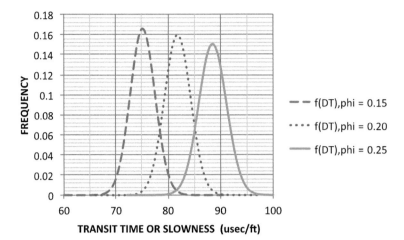

Figure 6.2 Probability density functions for ΔT for three different porosities. ΔT is trea-tedas a random variable because of the uncertainty in the parameters in the Wyllie equation.

$$\sigma_2 = (1-\phi)^2\sigma_1^2 + \phi^2\sigma_2^2 = 2.5^2(1-\phi)^2 + 7.5^2\phi^2 \qquad (6.3b)$$

Note that equation (6.3b) assumes that the random variables ΔT_{MA} and ΔT_F are uncorrelated.

Figure 6.2 plots the probability densities for the ΔT values for porosities 0.15, 0.20 and 0.25. With a difference in the porosities of 0.05, as shown here, it is evident that the distributions are largely separate, but do overlap. If ΔT was observed to have a value of 82 μs/ft, it is clear that the porosity would be about 0.20. However, for a transit time of 77 μs/ft it would be more difficult to be certain whether the porosity is 0.20 or 0.15.

Incidentally, once we have made the arbitrary decision to model ΔT with a normal distribution, the distribution of V_p cannot be normal and is given by:

$$f(V_p) = \frac{304.8}{\sqrt{2\pi\sigma_{DT}^2 V_P^2}}\exp\left[\frac{-\left(\dfrac{304.8}{V_P}-\mu_{DT}\right)^2}{2\sigma_{DT}^2}\right] \qquad (6.4)$$

if V_P is measured in km/s. (The factor 304.8 transforms μs/ft into km/s.) The non-normality of V_P is an artificial result, which follows from the assumption that ΔT_{MA} and ΔT_F are normal. This assumption is made purely for convenience, as it is much simpler to work with the slowness than with the velocity. For the sake of completeness, a sketch of the probability density function (equation 6.4) for V_P is shown in Figure 6.3.

The Wyllie equation can be used in other situations as well. For example, if well data are available we could construct an empirical correlation of ΔT against porosity, valid for that particular area (similar to the correlation represented by the thin line in Figure 6.1).

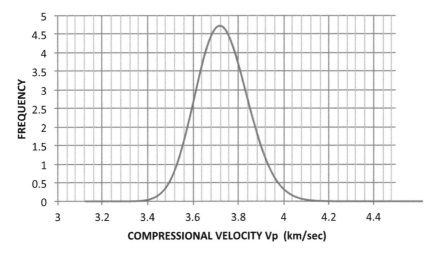

Figure 6.3 Probability density function for V_p for a porosity of 0.20, as given by equation (6.4). This distribution is not normal but clearly has a normal shape.

ESTIMATION OF THE SHEAR VELOCITY FROM THE COMPRESSIONAL VELOCITY

Castagna et al. (1985) have generated an empirical equation, which predicts the shear velocity V_s as a function of the compressional velocity V_p. An equation of this type can be very useful for forward modelling, and it is used extensively used in well log analysis. It is known as the "Mudrock" equation, and it applies to water-saturated sands and shale.

Figure 6.4 shows a crossplot of compressional velocity on the *x*-axis against shear velocity on the *y*-axis, corresponding to an interval about 1000 m thick. The fuchsia line is the Mudrock line, and it fits the data reasonably well.

Figure 6.4 illustrates how strongly the Castagna et al. (1985) correlation works, bearing in mind that the data come from a well taken at random. The small cluster of points in the NE corner of the plot, trending away from the Mudrock line, represents calcareous intercalations.

The equation for the Mudrock line is:

$$V_S = 0.862 V_P - 1.172 \tag{6.5}$$

where both velocities are measured in km/s. Despite the apparent robustness of the Mudrock line, we would like to estimate the uncertainties in the values of V_s calculated with equation (6.5). There are other correlations available which calculate the shear velocity as a function of the compressional velocity. One of them, due again to Castagna et al. (1993), is given by:

$$V_S = 0.804 V_P - 0.856 \tag{6.6}$$

(And there are others, which will not be considered here.)

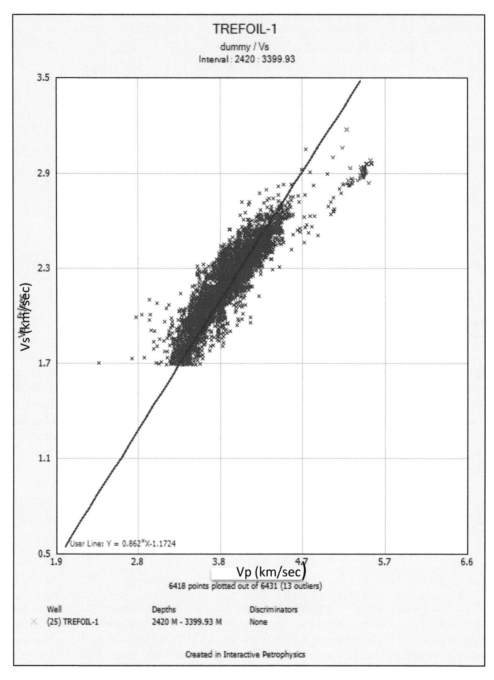

Figure 6.4 A crossplot with V_p on the x-axis and V_s on the y-axis. The Mudrock line is included in the plot.

Taking the average of equations (6.5) and (6.6) gives a third correlation:

$$V_S = 0.833V_P - 1.014 \tag{6.7}$$

which is the one we will use from now on. Suppose that the two coefficients in equation (6.7) are random variables and are normally distributed. That is, the factor multiplying V_p will be assumed to have a mean of 0.833 and a standard deviation chosen so there is a 95.5% chance that it lies between the coefficient 0.804 in equation (6.6) and the coefficient 0.862 in equation (6.5). The second random factor in equation (6.7) (the intercept term) will be assumed to have a mean of 1.014 and a standard deviation chosen so that 95.5% of its values lie between 0.856 and 1.172.

In other words, equation (6.7) is assumed to have the form

$$V_S = UV_P - W, \tag{6.8}$$

where U and W are normally distributed random variables whose properties are as described in the previous paragraph. The variable U is dimensionless, while the units of W are km/s. Given that U and W are normally distributed, V_S is also a normally distributed random variable. Table 6.1 lists the mean, variance and standard deviation of U and W, respectively:

The mean and variance of the random variable V_S are therefore

$$\mu = V_P\,\mu_U - \mu_W = 0.833V_P - 1.014 \tag{6.9}$$

$$\sigma^2 = V_P^2\sigma_U^2 + \sigma_W^2 = 2.1025 \times 10^{-4}V_P^2 + 0.6241 \times 10^{-2} \tag{6.10}$$

respectively. As before, the assumption that the shear velocity of the rock can be modelled as a normally distributed random variable has been made purely for the sake of mathematical convenience and is not intended to make any fundamental claim about the underlying rock physics.

Having introduced just two equations (Wyllie's and Castagna's), we can now calculate the compressional and shear velocities from a knowledge of the porosity, although with an associated uncertainty. Once the porosity is given, the rock density for a clean sandstone is easily calculated to be:

$$\rho_b = \rho_{MA}(1-\phi) + \rho_f\phi, \tag{6.11}$$

where the subscripts "MA" and "F" again refer to the densities of the matrix (quartz) and the fluid (water), respectively.

Table 6.1 Statistical properties of the random variables U and W

	Mean	St. dev	Variance
U	0.833	0.0145	2.1025 E(−4)
W	0.014	0.0790	0.6241 E(−2)

We can apply now the Monte Carlo method to estimate the distribution in the values of the shear and bulk moduli. Other methods for estimating these parameters are available, and the results of the simulation will be compared with these in order to check for consistency. We first list the input data for the simulation, then plot the resulting cumulative probability distribution for the shear modulus.

INPUT DATA FOR THE MONTE CARLO SIMULATION

Porosity: 0.20
Density of the fluid-saturated rock: 2.32 g/cc
Mean value of the slowness of the compressional wave: 81.8 µs/ft (see equation 6.3a)
Standard deviation of the slowness of the compressional wave: 2.5 µs/ft (see equation 6.3b)
Mean of random variable U: 0.833
Standard deviation of random variable U: 0.0145
Mean of random variable W: 1.014 km/s
Standard deviation of random variable W: 0.079 km/s

Figure 6.5 shows the simulation results for the shear modulus, with the modulus on the x-axis and the cumulative probability on the y-axis. The statistical properties of this distribution will be used later.

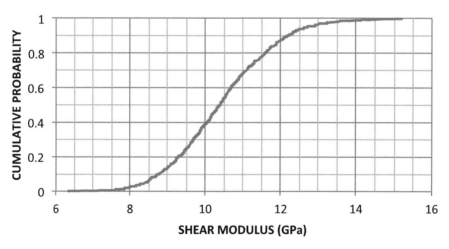

SHEAR MODULUS DISTRIBUTION FOR PHI=0.20

Figure 6.5 Cumulative probability distribution for the shear modulus. From the graph it is evident that there is about a 90% chance that the shear modulus of a clean sand with 20% porosity will be equal to or greater than 8.7 GPa. There is also a 90% chance that the shear modulus will be equal to or less than 12.2 GPa.

ESTIMATION OF THE ELASTIC PARAMETERS OF THE IDEAL ROCK USING THE HASHIN-SHTRIKMAN BOUNDS

In Chapter 5, we discussed in detail the Voigt-Reuss and Hashin-Shtrikman bounds, and demonstrated that the latter offers the least uncertainty when estimating parameters. Here we will estimate the values of the elastic parameters K_b and μ (the bulk and shear moduli, respectively), using the two Hashin-Strikman equations (6.13a) and (6.13b) from that chapter. The uncertainty in the values will be substantial, because the two bounds turn out to be quite far apart. But the bounds are useful, because they are not empirical formulas but instead have a solid theoretical basis grounded in the properties of isotropic elastic materials. Figure 6.6 plots the upper and lower Hashin-Strikman bounds for K_b, plus their average, as a function of porosity.

To estimate the uncertainty in K_b, we will again assume that the parameter can be regarded as a random variable, with a mean given by the average of the upper and lower bounds, and a standard deviation chosen that there is a 99.74% chance that the values of K_b lie between the upper and lower bounds.

That is, for any given value of the porosity, K_b can be regarded as a random variable, whose mean and variance depend on the upper and lower bounds and their average for that particular porosity.

We will now check whether our estimate of the distribution for K_b following from the Wyllie and Castagna equations is similar to the distribution of K_b values generated by the Hashin-Shtrikman bounds.

For a porosity of 0.20, the Wyllie-Castagna approach predicts a mean for K_b of 18.894 GPa and a standard deviation of 1.48875 GPa. Applying the Hashin-Shtrikman

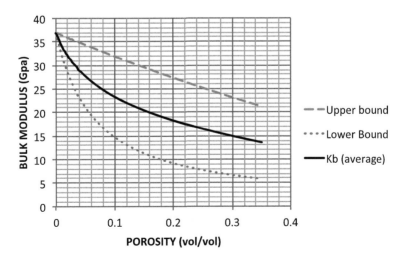

Figure 6.6 Estimate of the value of the bulk modulus or incompressibility K_b as the average of the lower and upper Hashin-Shtrikman bounds. As can be seen, the uncertainty is considerable. For instance, for a porosity of 0.20, the true value of K_b could be anywhere between about 9 and 27 GPa.

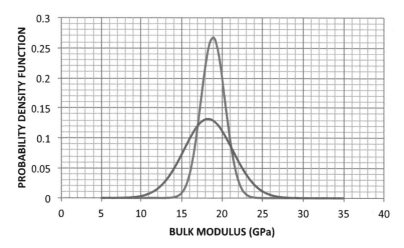

Figure 6.7 The two density functions for the values of K_b are derived from the Hashin-Shtrikman bounds (the flatter curve) and from the Wyllie and Castagna equations. The rock has a porosity of 0.2.

bounds instead gives a mean for K_b of 18.275 GPa and a standard deviation of 3.022 GPa. Note that the means in both distributions are very similar.

Figure 6.7 plots the probability density functions for both distributions. The curve with the higher frequency of values at its peak and a correspondingly lower variance is the density function for the values derived from the Wyllie and Castagna equations. The second curve corresponds to the average of the Hashin-Shtrikman bounds.

In principle, these results are rather satisfactory, particularly due to the closeness of the means. But in fact Figure 6.7 presents us with a problem that is common in probability analysis, where there are two different sources of information about the same variable and we would like to know which of the sources, if any, should be given more weight. If it is assumed that the two distributions are completely independent, we can combine them into a single distribution that is normally distributed and has a mean and variance given by:

$$\mu = \frac{\mu_1 \sigma_2^2 + \mu_2 \sigma_1^2}{\sigma_2^2 + \sigma_1^2} \tag{6.12a}$$

$$\sigma^2 = \frac{\sigma_1^2 + \sigma_2^2}{\sigma_1^2 \sigma_2^2} \tag{6.12b}$$

respectively. Figure 6.8 shows the two distributions from Figure 6.7, plus the overall distribution constructed by combining both sources of data.

Finally, we can disregard for the moment the fact that K_b is being treated as a random variable and compare its values over the range of porosities, assuming that all the variables involved take on their mean values. Figure 6.9 compares the K_b values calculated in this way from the Hashin-Shtrikman bounds and from the Wyllie and

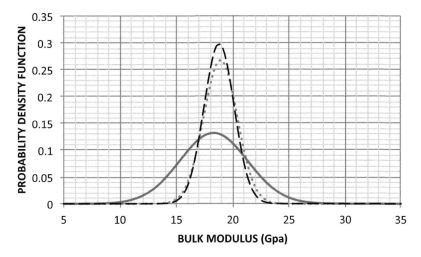

Figure 6.8 Same as Figure 6.7, but including the overall distribution (the dashed curve) found by combining the two sources of data. (From the Wyllie and Castagna equations and the Hashin-Shtrikman bounds.)

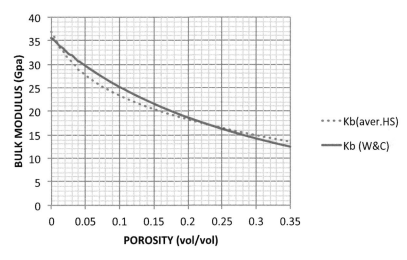

Figure 6.9 The two sets of average K_b values, calculated from the HS bounds and the Wyllie and Castagna equations, respectively. The two curves are close, which adds credibility to both methods.

Castagna equations. The two curves look very similar, and so the two independent methods yield comparable results. This is very satisfactory and adds credibility to both methods.

A similar analysis can be performed for the shear modulus. Figure 6.10 plots the upper bound and the average between the bounds for the shear modulus as a function of porosity. (Note that the lower bound for the shear modulus is zero for all values of the porosity.)

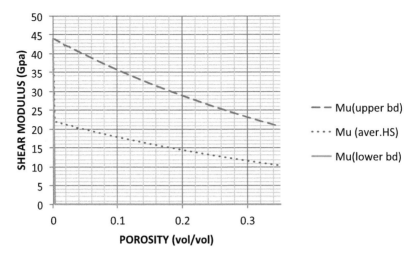

Figure 6.10 The upper and lower H-S bounds for the shear modulus and, their average, as a function of the porosity. The lower bound is zero for all values of the porosity, except when the porosity is equal to 0, where the lower bound "jumps" to a value of 44 GPa (the shear modulus of pure quartz). That is, the lower Hashin-Shtrikman bound is a discontinuous function at zero porosity.

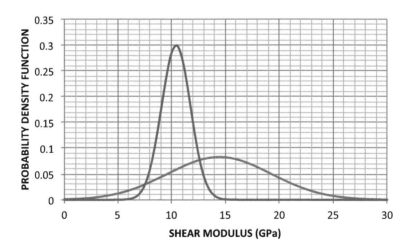

Figure 6.11 Probability density functions for the shear modulus calculated using the two methods, with a porosity of 0.2.

Figure 6.11 shows the probability density functions of the shear modulus as calculated from the Hashin-Shtrikman bounds and the Wyllie and Castagna equations, for a porosity of 0.2.

The curves appearing in Figure 6.11 are somewhat disappointing. The mean of the distribution generated from the Wyllie and Castagna equations is 10.43 GPa, whereas the mean calculated from the H-S bounds is 14.43. This is a large difference and does not validate the methods. One, or possibly both, of the two methods is probably wrong, but

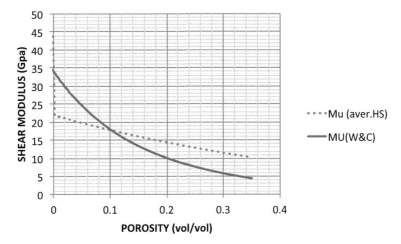

Figure 6.12 Comparison of the predicted values of the shear modulus values over the whole porosity range, assuming that all parameters have their mean values.

it is not possible to determine which. Thus, although the methods are consistent when calculating the bulk modulus, this consistency does not extend to the shear modulus.

Figure 6.12 confirms this inconsistency, as it demonstrates that over almost the whole range of porosity values the two methods predict substantially different values for the shear modulus, if all the parameters are assigned their mean values. This mismatch is unexpected and disappointing. In an attempt to determine which of the two methods gives the more accurate estimates of the shear modulus, we will introduce the results of three other methodologies and, finally, compare the estimates with real data.

CORRELATIONS USED TO ESTIMATE THE BULK AND SHEAR MODULI

Three alternative correlations that will be used for comparisons are the following

The Murphy et al. (1993) correlation

This correlation is valid only for clean sands (*i.e.* the correlation was generated exclusively from data for quartz sands):

$$K_{dry} = 38.18(1 - 3.39\phi + 1.95\phi^2)......\phi \leq 0.35$$

$$K_{dry} = \exp(-62.60\phi + 22.58)......\phi > 0.35$$

$$\mu = 42.65(1 - 3.48\phi + 2.19\phi^2)....\phi \leq 0.35$$

$$\mu = \exp(-62.69\phi + 22.73).....\phi > 0.35$$

The critical porosity hypothesis (Nur, 1992)

The critical porosity hypothesis assumes that there is a critical porosity above which the rock behaves like a suspension, meaning that the mineral grains are effectively suspended in the liquid. For sandstones, the critical porosity ϕ_C is estimated to be about 0.40. The correlation is then:

$$K_{dry} = K_0\left(1 - \frac{\phi}{\phi_C}\right) \qquad \mu = \mu_0\left(1 - \frac{\phi}{\phi_C}\right)$$

One problem with this particular approach is that the critical porosity of a sandstone could depend on the degree of shaliness and/or cementation.

The Krief et al. (1990) correlation

In principle, this correlation is valid for any lithology. It has the form:

$$K_{dry} = K_0(1 - \phi)^m .....m = \frac{3}{1 - \phi}$$

$$\mu = \mu_0(1 - \phi)^m .....m = \frac{3}{1 - \phi}$$

It should be noted that both Krief's correlation and the critical porosity hypothesis entail that:

$$\frac{K_{dry}}{\mu} = \frac{K_0}{\mu_0}$$

Figure 6.13 compares the calculated values of K_{dry} for clean sand as a function of porosity, using the Murphy and Krief correlations and the critical porosity hypothesis. It is evident that all three methods produce similar results.

In order to compare the results of the Hashin-Shtrikman and Wyllie and Castagna methods with the correlations, it is necessary to first use the Gassmann equation to transform the values of K_{dry} predicted by the correlations into values of the bulk modulus. The values predicted for the shear modulus can then be compared directly. Figure 6.14 makes such a comparison and plots the curves generated by the Hashin-Shtrikman, Wyllie & Castagna and Krief methods. It is evident that the three curves are completely inconsistent. Furthermore, except for porosities greater than about 0.20, the values predicted by the Krief method appear to be too high.

Figure 6.15 plots the bulk modulus as a function of the porosity, as calculated again from the Hashin-Shtrikman, Wyllie & Castagna and Krief methods (with the values of K_{dry} transformed into values of K_b using the Gassmann equation as before). The curve generated using the Krief method is now more consistent with the other two, and all three methods seem to be acceptable. So, in principle, one could use the Krief correlation together with the Gassmann equation to estimate the bulk modulus. However, the shear modulus remains problematic.

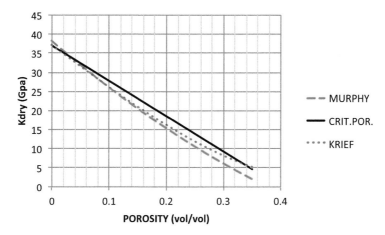

Figure 6.13 K_{dry} as a function of porosity for pure sands (K_0 = 37 GPa) using the methods of Murphy and Krief, and the critical porosity hypothesis. The three methods yield similar results, but only Krief's method can in principle be applied to any lithology.

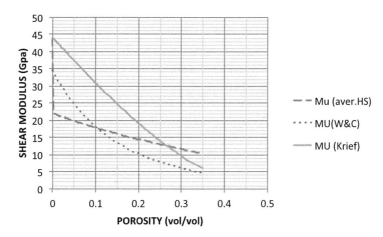

Figure 6.14 The values of the shear modulus calculated using the H-S, Wyllie & Castagna and Krief methods. The three curves are inconsistent and it is not possible to guess which is accurate. (Note that, as before, the H-S curve is discontinuous at zero porosity.)

COMPARISON WITH REAL DATA

The real data that will be used as a final test of the methods have been taken from a well with a complete set of logs. A 4.3 m interval that is believed to consist of a clean, water-bearing sand has been selected. The Gamma Ray values for this interval were very low, and there was almost no separation between the Neutron and Density logs. The actual bulk and shear moduli at every point in the interval were calculated using the DTCO, DTSM (compressional and shear slowness, respectively) and Density logs.

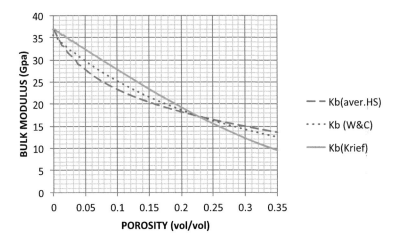

Figure 6.15 Values of the bulk modulus generated by the H-S, Wyllie & Castagna and Krief methods (the Krief correlation produces K_{dry}; it has been transformed to K_b using the Gassmann equation). The Krief values are consistent with the other two.

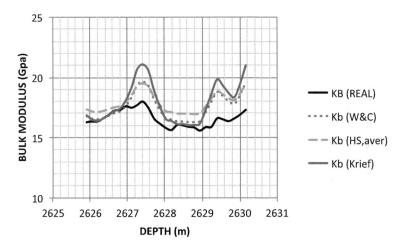

Figure 6.16 The bulk modulus as a function of depth for a very clean sand interval about 4.3 m thick. The actual values of the bulk modulus are the lower, thick continuous line. The other curves correspond to the values estimated using the Wyllie & Castagna equations, the H-S average and the Krief correlation together with the Gassmann formula.

As the rock is presumed to be a clean sand, with a negligible amount of shale, the matrix density is 2.65 g/cc (the density of quartz) and the porosity can be calculated from equation (6.11).

Figure 6.16 plots four sets of values of the bulk modulus as a function of depth, over the 4.3 m interval. The curves are: (a) the actual values of K_b; (b) the values of K_b calculated using the Wyllie and Castagna equations (with the parameters assumed to

Table 6.2 Average values of K_b for each of the curves in Figure 6.17, together with the average absolute error and the percentage error (which is 100 × Aver. Abs. error/Average Actual K_b)

	Real	Wyll & Cast	HS (Aver)	Gass
Average K_b	16.57	17.51	17.89	17.92
Aver Abs err		0.97	1.32	1.36
% Error		5.85	7.98	8.18

have their mean values); (c) the values of K_b derived from the H-S bounds; and (d) the values of K_b generated using the Krief correlation and the Gassmann formula.

Table 6.2 lists the values of some indicators of the goodness-of-fit of the three estimates of K_b.

It appears that the Wyllie-Castagna combination produces the best results, but for all three methods there are subintervals where the error is substantial. However, if the average errors only are considered, the three methods produce acceptable results. Note that all three methods are biased, in that the predicted values of K_b at every depth are greater than the actual values.

A similar analysis can be performed for the shear modulus over the same interval.

Figure 6.17 shows the actual and the three estimated values of the shear modulus as a function of depth. Note that, in this case, the only curve that fits the actual data reasonably well is the one constructed using the Wyllie and Castagna formulas. Table 6.3 lists the values of some parameters related to the goodness-of-fit for the different methods.

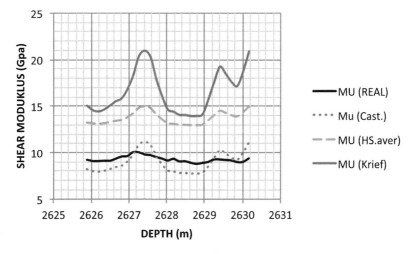

Figure 6.17 Shear modulus vs. depth for the actual data (in black) and the values estimated using the three methods discussed in the text. The only method to give satisfactory results is the Wyllie-Castagna method.

Table 6.3 Average values of the shear modulus for the actual data and the three estimation methods, together with the absolute error and the percentage error

	Real	Wyll & Cast	HS (Aver)	Krief & Gass
Average Mu	9.28	8.93	13.71	16.54
Aver Abs err		0.90	4.44	7.26
% Error		9.71	47.84	78.29

In the case of the bulk modulus, the advantage of using the Wyllie & Castagna method is slight, but for the shear modulus it is much greater. The strong bias in the values of the shear modulus generated using the H-S bounds and the Krief formula is evident in Figure 6.17.

SUMMARY

No theoretical formula exists that allows us to estimate the elastic constants of a clean sand with a known porosity. Instead, empirical formulas need to be used, and the accuracy of these empirical formulas is not guaranteed.

Three alternative methods for estimating the bulk and shear moduli have been investigated here:

a. Using the Wyllie and Castagna formulas, which produced the smallest errors in the estimates of the bulk and shear moduli.
b. Taking the average of the Hashin-Shtrikman bounds
c. Using the Krief correlation to calculate the shear modulus and K_{dry}, and then the Gassmann equation to estimate the bulk modulus.

A Monte Carlo simulation was also performed to estimate the uncertainty in the values of the shear modulus calculated using the first method.

The three methods produce consistent estimates for K_b in ideal rocks. In at least one case using real data, the results of the three methods were acceptable, but the Wyllie-Castagna approach tends to give the best results.

In the case of ideal rocks, the three methods are not consistent when used to calculate the shear modulus, and only the Wyllie-Castagna approach was reasonably accurate when compared with real data.

When performing an exercise in forward modelling for clean sands with a known porosity, the recommended approach therefore is to use the average of the Wyllie and Castagna equations as shown in equation (6.7).

It is notable that the Krief correlation produces very high values for the shear modulus in both ideal rock and in an actual rock sequence.

It is not clear why the bulk modulus can be predicted reasonably well using any of the three methods, whereas the shear modulus can be predicted with tolerable accuracy only with the Wyllie-Castagna formulas. The obvious conclusion to draw is that the real situation is much more complex than the models we are working with.

REFERENCES

Castagna, J., Batzle, M. and Eastwood, R. (1985), Relationships between compressional-wave and shear-wave velocities in clastic silicate rocks, *Geophysics*, 50, 571–581.

Castagna, J., Batzle, M. and Kan, T. (1993), Rock physics – The link between rock properties and AVO response, in Offset Dependent Reflectivity – Theory and practice of AVO analysis, J.P. Castagna and M. Backus, eds., *Investigations in Geophysics*, No.8, SPE, Tulsa, Oklahoma, 135–171.

Krief, M., Garat, J., Stellingwerff, J. and Ventre, J. (1990), A petrophysical interpretation using the velocities of P and S waves, *The Log Analyst*, 31, 355–369.

Mavko, G., Mukerji, T. and Dvorkin, J. (1998), *The Rock Physics Handbook*, Cambridge University Press, Cambridge-New York-Melbourne.

Murphy, W., Reischer, A. and Hsu, K, (1993), Modulus decomposition of compressional and shear velocities in sand bodies, *Geophyusics*, 58, 227–239.

Nur, A. (1992), Critical porosity and the seismic velocities in rocks: EOS., *Transactions of the American Geophysical Union*, 73, 43–66.

Wyllie, M., Gregiry, A. and Gardner, L. (1958), An experimental investigation of factors affecting elastic waves velocities in porous media, *Geophysics*, 23, 459–493.

APPENDIX 6.I: ESTIMATION OF THE INCOMPRESSIBILITY OF THE SOLID PART OF THE ROCK IN SHALEY SANDS

If forward modelling or a fluid substitution exercise is performed on shaley sand (or a sandy shale), the Wyllie formula is no longer applicable, because of the difficulty in determining the value of ΔT_{MA} (although the Castagna equation is still applicable). In practice, forward modelling is not possible for shaley sand if there are no wells in the vicinity of the location where we want to drill. Well data are essential in this case for the evaluation of certain critical parameters.

What follows is an exercise performed on data from a real well to estimate the incompressibility of the solid part of the rock, as a function of depth.

The well log data are taken from a 545 m interval containing a sand/shale sequence. The values of PHIE, V_{sh} and S_{we} are available for this sequence, as well as the bulk and shear moduli, as the density and the primary and shear velocity logs are included.

Figure 6.1.1 shows a histogram of the values of the bulk modulus for points in the sequence where V_{sh} is greater than 0.95. The average of all these points is about 22.1 GPa, and we will regard this as the value of the incompressibility K_{sh} of the shale. Although not shown here, a similar histogram has been prepared for the shear modulus at all points where $V_{sh} > 0.95$, and the mean of these values is 10.75 GPa. This will be taken to be the shear modulus μ_{sh} of the shale.

The solid part of the rock will be modelled as a mixture of sand and shale, where as usual the quartz has an incompressibility of 37 GPa, and the shale is assumed to have an incompressibility of 22.1 GPa. To estimate the incompressibility K_0 of the solid part of the rock, we calculate the Hashin-Shtrikman upper and lower bounds, then average them. The shear modulus of the solid part of the rock is estimated in the same way. Because each point in the column has a different value of V_{sh} and effective porosity, we will estimate the parameters of the solid part of the rock as functions of $X_{sh} = V_{sh}/(V_{sh} + V_{qtz})$, the volume fraction of shale in the solid part of the rock. Note that $V_{sh} + V_{qtz}$ is equal to 1 minus the effective porosity. The shear modulus of the solid part of the rock at each point in the column can be estimated using a similar approach.

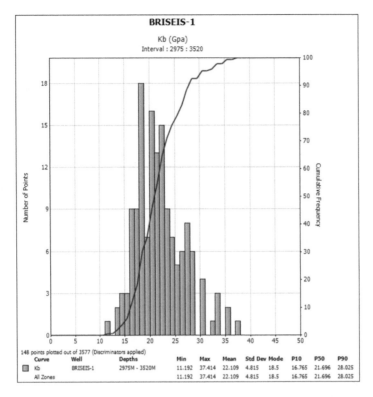

Figure 6.1.1 Estimation of K_{sh}. The histogram plots the frequencies of the values of K_b from all points where V_{sh} is greater than 0.95. The average of these values is taken to be the incompressibility K_{sh} of the shale.

$$K_{0(UPPER)} = K_{qtz} + \cfrac{X_{sh}}{\left(K_{sh} - K_{qtz}\right)^{-1} + \left(1 - X_{sh}\right)\left(K_{qtz} + \dfrac{4}{3}\mu_{qtz}\right)^{-1}}$$

$$K_{0(LOWER)} = K_{sh} + \cfrac{\left(1 - X_{sh}\right)}{\left(K_{qtz} - K_{sh}\right)^{-1} + X_{Ssh}\left(K_{sh} + \dfrac{4}{3}\mu_{Ssh}\right)^{-1}}$$

The "true" value of K_0 will be taken to be the average of the upper and lower bounds, which are calculated using the following equations:

Figure 6.1.2 plots the upper and lower bounds and their average as functions of X_{sh}, the shale volume fraction in the solid part of the rock. It is clear from Figure 6.1.2 that the difference between the upper and lower bounds is relatively small in this case, so the uncertainty in the value of K_0 is acceptably low.

We can follow an analogous procedure to estimate the shear modulus of the solid part of the rock, where the upper and lower bounds to be averaged are:

$$\mu_{0UPPER} = \mu_{QTZ} + \cfrac{X_{SH}}{\left(\mu_{SH} - \mu_{QTZ}\right)^{-1} + \cfrac{2\left(1 - X_{SH}\right)\left(K_{QTZ} + 2\mu_{QTZ}\right)}{5\mu_{QTZ}\left(K_{QTZ} + \cfrac{4}{3}\mu_{QTZ}\right)}}$$

and

$$\mu_{0LOWER} = \mu_{SH} + \cfrac{\left(1 - X_{SH}\right)}{\left(\mu_{QTZ} - \mu_{SH}\right)^{-1} + \cfrac{2X_{SH}\left(K_{SH} + 2\mu_{SH}\right)}{5\mu_{SH}\left(K_{SH} + \cfrac{4}{3}\mu_{SH}\right)}}$$

Figure 6.1.2 is a crossplot of the incompressibility of the solid part of the rock on the y-axis against the volume fraction of shale in the matrix or solid part of the rock on the x-axis. At any point where the effective porosity and the shale volume fraction are known, we can calculate K_0 in this way, although with an inherent uncertainty. This uncertainty derives from the fact that our value of K_0 is an average of two bounds, and so the true value may lie anywhere between the bounds, while the parameter K_{sh} has a substantial spread, as is evident in Figure 6.1.1.

Although we saw earlier that the Krief correlation can be problematic, we will again test the reliability of the combination of the Krief and Gassmann equations using well log data from a column containing sands and shaley sands. Once the value of K_0 is known, K_{dry} and the shear modulus can be estimated at each point in the column, and the Gassmann equation can be applied to transform K_{dry} into K_b. The values of μ and K_b calculated in this way can then be compared with the values derived from the Velocity and Density logs.

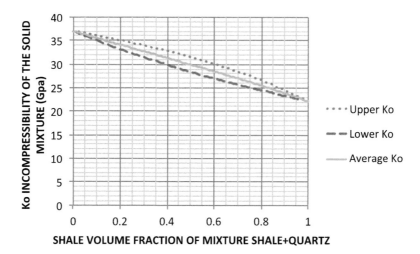

Figure 6.1.2 Incompressibility of the solid part of the rock (K_0) for mixtures of quartz and shale. The maximum uncertainty in the true value of K_0 occurs when $x = 0.52$. At this point, the uncertainty in K_0 is 29.58 ± 1.59 GPa.

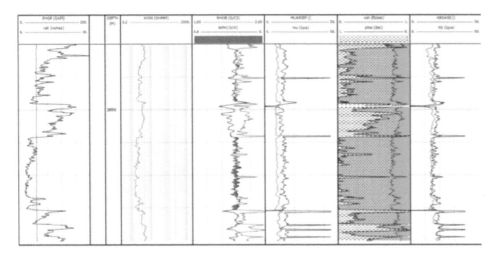

Figure 6.1.3 Well log data for an interval 60 m thick are used to test the reliability of the Krief correlation.

Figure 6.1.3 shows a portion of the well log data with which the analysis was carried out and the final results. The interval is 60 m thick.

The leftmost track shows the Gamma Ray log, followed by a column listing the measured depth (which is empty) and then the TVD depth track. (Note that the column is vertical.) Next to the TVD track is the resistivity track, followed by the Neutron/Density track. The Neutron and Density logs have been used to calculate V_{sh} and PHIE. The Gamma Ray log was also used to calculate V_{sh} over a gas-bearing interval (which is not shown in the figure). The next track (third from the right) plots the values of the shear modulus calculated using the Krief correlation (in black), together with the actual values of the shear modulus, calculated from the shear sonic velocity and the density. It is this comparison that is the focus of the exercise. The shear modulus track is followed by a graphic track (second from the right), displaying the porosity and shale volume fraction. Finally, the rightmost track contains two curves: one showing the bulk modulus calculated using the Krief and Gassmann equations (in red) and the other the actual bulk modulus, calculated from the shear velocity, primary velocity and density logs.

It should be noted that the predicted values of the bulk modulus are largely accurate (see the rightmost track in Figure 6.1.3). However, the accuracy of the predictions was expected to be better, particularly over the very clean sands, which lie between approximately 3060 and 3080 m (just below the 3050 m mark in Figure 6.1.3). In clean sands, the Hashin-Shtrikman bounds are not needed to estimate the incompressibility of the solid part of the rock, so we can avoid the introduction of somewhat dubious parameters such as K_{sh} and μ_h. (The incompressibility of the solid part of the rock in clean sands like these should be close to 37 GPa.)

The predicted values of the shear modulus, in the third track from the right, lie systematically above the actual values, and the errors in these estimates might not be considered acceptable.

Chapter 7

Applications of rock physics to AVO analyses

REFLECTION COEFFICIENTS

If a seismic wave is incident at right angles to a surface separating two isotropic media with different elastic properties, a part of the energy will be transmitted and a part of it will be reflected. If the upper medium is labelled "1" and the lower medium "2", the reflection coefficient R_P for the P-waves is defined to be:

$$R_P = \frac{Z_2 - Z_1}{Z_2 + Z_1} = \frac{\rho_2 V_{P2} - \rho_1 V_{P1}}{\rho_2 V_{P2} + \rho_1 V_{P1}} \tag{7.1}$$

where Z_2 is the acoustic impedance of the lower medium and Z_1 is the impedance of the upper medium. As equation (7.1) indicates, the impedance is the product of the P-velocity V_P in the medium with its density ρ. An analogous equation can be written, for instance, for the S-velocity.

In very favourable circumstances, there can be a direct correlation between the reflection coefficients and the observed seismic amplitudes in a seismic trace. Furthermore, the magnitudes of the reflection coefficients – and hence of the amplitudes – change with the angle of incidence of the seismic wave.

Figure 7.1 shows what is meant by the angle of incidence of the seismic wave. It shows a section containing a horizontal interface that separates two media with different elastic properties. The point labelled "source" represents a shot point, and the "receiver" point is the location of a geophone. The wave is reflected by the interface and arrives at the receiver with an angle of incidence equal to θ. The conventional AVO [Amplitude Variations with Offset] technique makes use of a number of geophones. Each geophone receives information from waves reflected with a different angle from the reflecting surface.

The change in the magnitude of the reflection coefficients with the angle of incidence is the basis of AVO. These changes are a function of the elastic properties of the two media on either side of the reflection interface. Zoeppritz (1919) first developed a set of equations to describe the variation of the coefficients with offset in the general case (not just for the P-impedance). However, the Zoeppritz equations are cumbersome and difficult to handle. We will work instead with an approximation to the Zoeppritz equations generated by Aki and Richards (1980). In order to derive this approximation, these authors assumed that there is a small contrast in the properties of the two layers separated by the interface. They also assumed that the maximum angle of incidence is

DOI: 10.1201/9781003261773-8

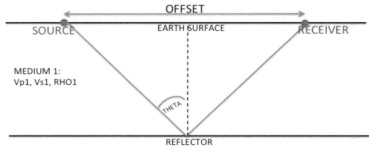

Figure 7.1 A simple sketch illustrating the angle of incidence for one receiver.

less than the critical angle, which is the angle of incidence of a wave which after transmission runs parallel to the interface (meaning that, in the language of Snell's law, the angle of refraction is equal to 90 degrees). The Aki-Richards approximation is, in fact, accurate up to angles of incidence of about 40°.

The Aki-Richards approximation formula is as follows:

$$R_P \approx R_{P0} + B\sin^2(\theta) + C\left[\tan^2(\theta) - \sin^2(\theta)\right], \tag{7.2}$$

where

$$R_{P0} \approx \frac{1}{2}\left(\frac{\Delta V_P}{V_P} + \frac{\Delta \rho}{\rho}\right) \tag{7.3}$$

is the reflection coefficient at normal incidence,

$$B = \frac{1}{2}\frac{\Delta V_P}{V_P} - 2\left(\frac{V_S}{V_P}\right)^2\left(\frac{\Delta \rho}{\rho} + 2\frac{\Delta V_S}{V_S}\right) \tag{7.4}$$

and

$$C = \frac{1}{2}\frac{\Delta V_P}{V_P}, \tag{7.5}$$

while the various symbols appearing in equations (7.3), (7.4) and (7.5) are defined by

$$\Delta\rho = \rho_2 - \rho_1$$

$$\Delta V_P = V_{P2} - V_{P1}$$

$$\Delta V_S = V_{S2} - V_{S1}$$

$$\rho = \frac{\rho_1 + \rho_2}{2}$$

$$V_P = \frac{V_{P1} + V_{P2}}{2}$$

$$V_S = \frac{V_{S1} + V_{S2}}{2}$$

The reflection coefficient at normal incidence, R_{P0}, is also called the intercept. The coefficient B defined by equation (7.4) is known as the gradient, while the coefficient C defined by equation (7.5) is called the curvature. The values of the intercept and the gradient calculated from seismic data are often used to identify interfaces of interest (such as, for example, a shale overlying a gas sand).

Figure 7.2 plots the magnitude of the P-wave reflection coefficient as a function of the angle of incidence for a particular gather and reflecting horizon. This graph was constructed using equation (7.2) and represents an interface whose upper boundary is a shale and lower boundary is a gas-bearing sand. The values of the velocities and the density at the boundary were taken from real data. The reflection coefficient corresponding to zero angle of incidence is the intercept, and in this case is negative, while the gradient B is negative because the slope of the curve is also negative. If a plot of gradient against intercept were plotted for a range of different horizons and gathers, the point corresponding to Figure 7.2, with a negative intercept and a negative gradient, would indicate a so-called Class III anomaly. In general, Class III anomalies tend to be produced by an interface where shale overlies a gas sand, but even in cases where the seismic data are similar to Figure 7.2 (which models a situation with

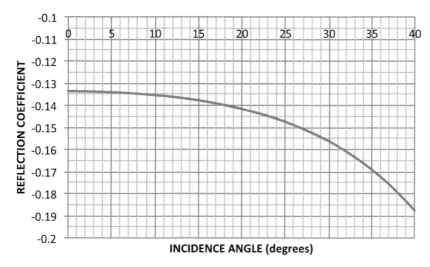

Figure 7.2 The reflection coefficient as a function of the angle of incidence, calculated from equation (7.2) for one particular horizon. In this case, both the intercept and the gradient are negative, a fact which would normally be interpreted as evidence of an interface where shale overlies a gas sand.

a large contrast between the physical properties of the shale and the gas sand) there is no guarantee that the anomaly is produced by an interface of this type. We will check later whether, for a particular interface, the intercept and gradient remain constant as the depth changes.

The intercept and gradient can be calculated using seismic data alone, and there is no need for extra well data. However, if only seismic data are available, the information that can be extracted using the AVO technique will be limited. If wells are present in the area, on the other hand, it might be possible to estimate the velocities and densities of all the major lithological types in the formation. The various types of interfaces to be expected can then be simulated and modelled as points on a crossplot of intercept against gradient, so that a comparison with the actual seismic data can be made. Figure 7.3 shows a crossplot of simulated values of the intercept and gradient, calculated on the basis of the physical properties of the various lithological types present in the wells close to a particular area of study. The methods used to generate the simulated values will be described in more detail shortly.

It should be stressed that the formulas presented above are valid only in cases where both media are isotropic. If at least one of the media is anisotropic, the equations need to be modified (Rüger, 2002). Throughout this chapter, it is assumed that both media are isotropic, and so the Aki-Richards approximation can be applied.

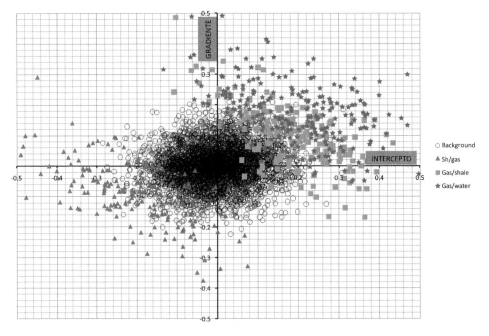

Figure 7.3 Simulated values of intercept and gradient for depths of about 1000 m below the sea floor, corresponding to interfaces of interest that would be expected given the information from nearby wells. The red circles represent interfaces with shale overlying gas sand. Most of them have a negative intercept and a negative gradient, corresponding to a Class III anomaly. The dark blue circles represent interfaces with gas sand overlying water sand.

SIMULATION OF THE AVO RESPONSES

If there is a well present near the area of study, the physical properties of all the lithological types in the well can be determined as functions of depth. Figure 7.4 shows three crossplots, with the depth below the sea floor along the *x*-axis, and (from left to right) the P-velocity, the S-velocity and the density along the *y*-axis, for rocks in the sequence that have been identified as shale. Note that there is a good correlation between the three physical properties and the depth, as there should be. No abnormal pressures have been detected at the range of depths shown in the plots, so the porosity (which is strongly tied to these properties) should also be closely correlated with the depth in the well. In order for the well data to be useful, the area where the AVO study is being performed should, of course, have similar lithology and similar pressure conditions.

Figure 7.5 is the same type of diagram as Figure 7.4, but with the data taken from gas-bearing sands. In fact, at only a couple of points in the well sequence were the rocks identified as gas-bearing sands, so the remaining data points have been calculated using the Gassmann equation.

A similar plot, not shown here, has been made for water sands. Although there was no oil in this particular well, the Gassmann equation was also used to generate the corresponding plot for oil sands. It should be noted that substantial bentonite layers were also encountered in the well, and their properties have also been estimated for inclusion in the model.

Using this data, it is possible to get a quick impression of the type of normal incidence reflection coefficients that would be expected for the interfaces of interest. Figure 7.6 shows a crossplot of the P-impedance against the depth below the sea floor, based on the information furnished by the well. At this stage of the analysis, it is assumed that the three basic parameters (the P-velocity, S-velocity and density) are deterministic functions of the depth, with their values fixed by the lines of best fit shown in Figures 7.4 and 7.5, and the analogous figures for water sands, oils sands and bentonite.

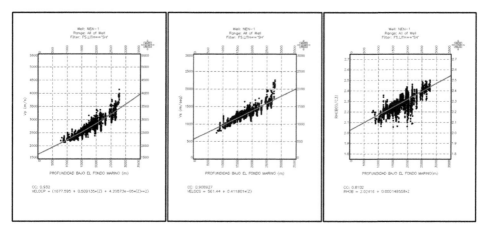

Figure 7.4 P-velocity (left), S-velocity (centre) and density (right) in a shale sequence, plotted as functions of the depth below the sea floor. Note the good correlation between these three properties and the depth.

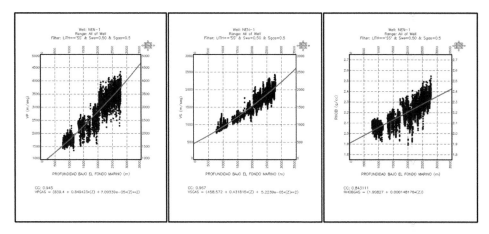

Figure 7.5 P-velocity (left), S-velocity (centre) and density (right) for gas sands, plotted as functions of the depth below the sea floor. There were few sands in the well from which Figure 7.4 was constructed that were actually gas-bearing. The responses of the remaining gas sands in this figure were calculated using the Gassmann equation.

Consider the interface between an upper shale and a lower gas sand. At a depth of 1000 mbsf, according to Figure 7.6, the gas sand impedance (the dotted line) will be lower than the shale impedance (the dashes & dots line), and therefore, the normal incidence reflection coefficient will be negative (*cf.* equation 7.1). That is, this interface will produce a negative intercept. At greater depths, the contrast between the two impedances decreases, and the magnitude of the (negative) normal incidence reflection coefficient will become smaller. At about 2500 mbsf, the contrast disappears, and the reflection coefficient should be zero. Below this depth, the contrast begins to increase, but now the impedance of the gas sand is greater than the impedance of the shale, which entails that the normal incidence reflection coefficient (or intercept) is positive. The lesson to be drawn from this is that, at shallow depths, an interface where shale overlies a gas sand should indeed generate a Class III anomaly (assuming that the gradient is also negative), but in the case shown here it no longer gives rise to a Class III anomaly below 2500 mbsf.

Another point of interest is that, at shallow depths, the water sands and the shale are almost indistinguishable, with a normal incidence reflection coefficient close to 0. But at depths below 2500 mbsf, an interface with shale above a water sand will produce an intercept that is noticeably positive. On the basis of these quick observations, we can speculate that an AVO study would be more effective at shallow depths, where there is a substantial separation between the impedances of the shale and the gas-bearing sands, and the corresponding normal incidence reflection coefficients are strongly negative. As would be expected, and is confirmed in Figure 7.6, the contrasts at interfaces where a shale overlies an oil sand are less conspicuous.

However, in the range from about 2000 to 2500 mbsf, all the impedances curves (except the bentonite curve) are close together, and the normal incidence reflection coefficients become very weak and therefore difficult to identify. Even in the absence

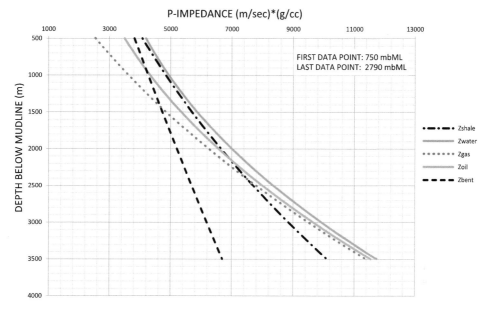

Figure 7.6 P-Impedance as a function of depth for shale, bentonite and water-, oil- and gas-bearing sands in a real well. The observations cover the range from 750 to 2790 mbsf.

of information about the gradients, it is clear that, in principle, an AVO study would be more effective for relatively shallow depths.

As stated earlier, the purpose of generating a crossplot of simulated intercepts and gradients, such as in Figure 7.3, is to compare the simulated data with the actual intercepts and gradients estimated from the seismic data. This is despite the fact that the simulated data correspond to properties of the reflection coefficients, whereas the actual intercept-gradient curve is generated from seismic amplitudes. In general, a rough correlation between the reflection coefficients and the amplitudes is to be expected, but depending on the geological conditions, they might be very different. This problem will be discussed briefly later on.

At this point, we will give a detailed description of the methods used to produce a diagram such as Figure 7.3. The entire process can be carried out using Excel and is easy to replicate. A plot of this type may be helpful for identifying AVO anomalies across a certain range of depths, when compared with gradient versus intercept plots derived from the seismic data. Although the computational effort needed to create these plots does not necessarily guarantee accurate results, particularly for depths below 1500–2000 mbsf, the authors nonetheless believe that this effort is worthwhile.

We will assume, as before, that well data exist from rock sequences not far from the area of study and that the lithological types present in the wells have been identified. In clastic sequences, of course, the rock types are predominantly sand and shale. The first step is to estimate the values of the three basic parameters (the P-velocity V_P, the S-velocity V_S and the bulk density ρ_b) as functions of the depth below the sea floor, for all the relevant lithological types – recalling that gas-bearing sands and water sands are distinct types. These functions are taken from the least-squares fit to the data, as in Figures 7.4 and 7.5, so the standard error of the regression should be available for each of them. The correlation coefficients between V_P and V_S, ρ_b and V_P, and ρ_b and V_S are also needed.

Among the sands, the curve for water sands can be calculated directly from the data, but in most cases the curves for the gas- and oil-bearing sands need to be constructed using the Gassmann equation. We can presume that the proportion of sand and shale in the area of study is the same as the proportion observed in the nearest well. The relative proportions of water, oil and gas sands can be chosen arbitrarily (although the water sands should be more abundant than the oil- and gas-bearing sands).

Table 7.1 contains all the information needed to carry out the simulation exercise. We will work with a model abstracted from a real case, where only three lithological types are present: shale, water sands and gas sands. The proportions are chosen arbitrarily to be 0.6 shale, 0.3 water sands and 0.1 gas sands. We are particularly interested in the properties at points corresponding to an interface between an upper shale and a lower gas sand, and will analyze a section ranging between 500 and 1000 mbsf.

At each depth, the lithological type above the interface is selected at random, using the following procedure. A random number n is generated from the uniform distribution on [0, 1]. If $n < 0.6$ the lithology is shale, if $0.6 < n < 0.9$ it is a water sand, and if $n > 0.9$ the lithology is a gas sand.

At each depth and for each lithological type, the three basic parameters (the P-velocity, the S-velocity and the bulk density) are treated as random variables. The means of the parameters are calculated from the regression equations shown in the second column of Table 7.1. The values of the parameters themselves are generated randomly from normal distributions with means and standard deviations calculated

Table 7.1 The information needed to generate an AVO simulation.

Shale	F(depth)	Std. Error	Correl V_p	Correl V_s	Correl R_{hob}
V_p	$1677.595 + 0.509135D + 4.2087E(-05)\,D^2$	110	1	0.901	0.773
V_s	$561.44 + 0.4118D$	73	0.901	1	0.75
R_{hob}	$2.0246 + 0.00048558D$	0.15	0.773	0.75	1
Gas sand					
V_p	$839.4 + 0.8494D + 7.093E(-05)\,D^2$	100	1	0.887	0.711
V_s	$485.57 + 0.431815D + 5.223E(-50)\,D^2$	50	0.887	1	0.69
R_{hob}	$1.90827 + 0.000148178D$	0.2	0.711	0.69	1
Water sand					
V_p	$1420 + 0.7333D + 5E(-05)\,D^2$	89	1	0.91	0.761
V_s	$439.85 + 0.4166D + 6E(-05)\,D^2$	66	0.91	1	0.73
R_{hob}	$2.081875 + 0.00011325D$	0.17	0.761	0.73	1

Data for three rock types are listed: shale, gas sands and water sands. The column labelled "f(depth)" presents the regression equations linking the three physical properties (compressional and shear wave velocities and the density) with the depth below the sea floor, for the three rock types. The column labelled "Std. Error" lists the standard error of the estimate for each regression curve. The three rightmost columns contain the pairwise correlation coefficients between the three variables for each of the rock types.

using the parameters means, and the standard errors in the regression and correlation coefficients taken from the four rightmost columns in Table 7.1.

To illustrate this procedure, let us first generate a random value for the P-wave velocity, for a particular depth. To do this, we generate a random number between 0 and 1, which we label "probability", then use the following Excel command to generate a value for V_P:

=INV.NORMAL (probability, mean, standard deviation)

where the numbers "mean" and "standard deviation" are the uncorrected values of the V_P parameter mean and standard error in the regression, respectively.

The next step is to generate a value for V_S, the shear wave velocity, on the assumption that V_P and V_S are jointly distributed with a two-dimensional normal distribution. Because V_P and V_S are not independent random variables, the distribution for V_S alone will be affected by the value that has already been generated for V_P. Both the mean and the standard deviation for the distribution of V_S are now different from the uncorrected values. The conditional mean of V_S, given that $V_P = V_{Pk}$, is:

$$\mu[V_S | V_P] = \mu_{Vs} + \rho \frac{\sigma_{Vs}}{\sigma_{Vp}} \left(V_{Pk} - \mu_{VP} \right) \tag{7.6}$$

where μV_S is the unconditional mean of V_S, taken from the second column in Table 7.1, σV_S is the unconditional standard deviation of V_S, taken from the third column in Table 7.1, μ_{VP} and σV_P are the mean and standard deviation used previously when calculating the P-wave velocity, and ρ is the correlation coefficient between V_P and V_S, taken from the fourth column in Table 7.1.

The conditional standard deviation of V_S, given that $V_P = V_{Pk}$, is:

$$\sigma[V_S | V_P] = \sigma_{VS} \sqrt{1 - \rho^2} \tag{7.7}$$

To generate a value for V_P, we again generate a random number between 0 and 1, which we label as "probability", and use the Excel command

=INV.NORMAL (probability, mean, standard deviation)

where the numbers "mean" and "standard deviation" are now the conditional values calculated using equations (7.6) and (7.7), respectively.

By way of illustration, we will generate specific values for V_P and V_S. Assume that the first lithological type created is a gas sand. V_P is then treated as a normally distributed random variable with mean (in m/s) given by

$$\mu_{VP} = 839.4 + 0.8494\, D + 7.093 \times 10^{-5}\, D^2$$

(cf. Table 7.1) and a standard deviation of 100 m/s (see also Table 7.1). Similarly, the unconditional mean of V_P is given by

$$\mu_{VS} = 458.57 + 0.431815\, D + 5.223 \times 10^{-5}\, D^2$$

Table 7.2 Generation of random values for V_P and V_S at a depth of 521 m.

Depth	Random	Lith_up	V_p Average	Random	$V_p l$ (up)	V_s Average	Random	$V_s l$ (up)
521	0.932245291	Gas_ sand	1301.19071	0.3644086	1266.520776	682.338719	0.54459604	684.925026

The lithology above the interface is gas sand. Three of the columns contain random numbers. The first random number was used to generate the lithology, the second random number – to the left of the column marked "$V_p l$(up)" – is used to generate the P-wave velocity $V_p l$(up), and the third random number is used to generate a value $V_s l$(up) for the S-wave velocity.

and the unconditional standard deviation is 50 m/s. At a depth $D = 521$ m, the values we randomly generated for V_P and V_S are given in Table 7.2.

In Table 7.2, the column labelled "Vs. average" is the conditional average given by equation (7.6), and the standard deviation needed to generate the random value of V_S was calculated from equation (7.7).

The next step involves randomly generating a value for the bulk density ρ_b. The random variables V_P, V_S and ρ_b are assumed to be jointly distributed with a three-dimensional normal distribution, so the fact that both V_P and V_S have already been assigned specific values affects the conditional distribution for ρ_b.

At this point we will simplify the notation by labelling the three random variables X_1, X_2 and X_3 in place of V_P, V_S and ρ_b, respectively. The symbol x_1 then denotes a realization of the random variable X_1 and in general x_k denotes a realization of the kth random variable, while μ_k is the unconditional mean of X_k and σ_k is its unconditional standard deviation. Once X_1 and X_2 have been assigned specific values, the conditional mean of the random variable X_3 (in this case, the bulk density ρ_b) is given by:

$$\mu\left[X_3 \mid x_1, x_2 \right] = \mu_3 - \frac{c_{13}}{c_{33}}\left(x_1 - \mu_1 \right) - \frac{c_{23}}{c_{33}}\left(x_2 - \mu_3 \right), \tag{7.8}$$

where

$$\frac{c_{13}}{c_{33}} = \frac{\sigma(\rho_{12}\rho_{23} - \rho_{13})}{\sigma(1 - \rho_{12}^2)}$$
$$\frac{c_{23}}{c_{33}} = \frac{\sigma_3(\rho_{23} - \rho_{12}\rho_{13})}{\sigma_2(1 - \rho_{12}^2)}$$

Here, ρ_{12} represents the correlation coefficient between variables 1 and 2, ρ_{13} is the correlation coefficient between variables 1 and 3, and ρ_{23} is the correlation coefficient between variables 2 and 3.

The conditional variance of variable 3, once both X_1 and X_2 have been assigned specific values, is given by:

$$\sigma^2\left[X_3 \mid x_1, x_2 \right] = \sigma_3^2 \frac{\left[1 + 2\rho_{12}\rho_{13}\rho_{23} - \left(\rho_{12}^2 + \rho_{13}^2 + \rho_{23}^2 \right) \right]}{\left(1 - \rho_{12}^2 \right)} \tag{7.9}$$

Equations (7.8) and (7.9) are derived from the theory of the normal distributions in n dimensions. For more details, refer to Sveshnikov (1968).

All the numbers necessary for the calculations involved in equations (7.8) and (7.9) can be found in Table 7.1, while the realizations of variables 1 and 2 (V_P and V_S) were randomly chosen in the two previous steps.

Finally, to generate a value for ρ_b, we again choose a random number between 0 and 1 (the "probability" and call the command):

=INV.NORMAL (probability, mean, standard deviation)

where the number "mean" is the conditional value calculated using equation (7.8), and "standard deviation" is the square root of the conditional value calculated using equation (7.9).

Once we have generated the P-wave velocity, S-wave velocity and bulk density for the lithology above the interface, we repeat the procedure by randomly choosing a lithology below the interface and then generating the values of V_P, V_S and ρ_b at the same depth in the lower lithology. Once all six values are fixed, equations (7.3) and (7.4) can be applied to calculate the intercept and the gradient for this particular realization. The result is a single point in a diagram such as Figure 7.3. Of course, the process has to be repeated many times in order to generate a cloud of points.

SOME OF THE PROBLEMS IN USING AMPLITUDES AS SURROGATES FOR REFLECTION COEFFICIENTS

As a first approximation, the seismic amplitudes (*i.e.* the wiggles we see in a seismic section) are convolutions of wavelets with the reflection coefficients characteristic of the contrast between neighbouring lithological types within a rock body. The wavelet itself represents a moving pulse of energy, which is injected into the earth at the shot point during the course of a seismic survey. The mathematical details of the convolution process are described in many Geophysics textbooks. Al-Sadi (1980), for example, explains convolution as a graphical sliding operation, which is easy to apply in real cases.

Figure 7.7 plots the distribution of a large sample of simulated reflection coefficients against the one way time (OWT). The simulation models a section containing two thick shale bodies sandwiching a gas sand, whose location in OWT ranges between 1140 and 1146 ms. For a P-wave velocity of about 2500 m/s, this corresponds to a thickness of the gas sand in the order of 15 m. The reflection coefficients inside the two shale bodies, above and below the gas sand, are small because the physical properties change only slowly within the shale. The location of the gas sand can be identified immediately from the two large spikes in this plot, one on the left at 1140 ms and the other on the right at 1146 ms, which mark the upper and lower interfaces, respectively. The reflection coefficient at the upper interface is negative because the gas sand has a lower impedance than the shale.

Figure 7.8 plots the corresponding simulated amplitudes, superposed on the original reflection coefficients. The top and the base of the gas sand can still be identified from the amplitudes, although the locations of the interfaces are slightly displaced, and the apparent thickness of the gas sand is greater than the actual thickness. A bed of this type would show up as a Class III anomaly, at the right position in a gradient versus intercept diagram.

REFLECTION COEFFICIENTS

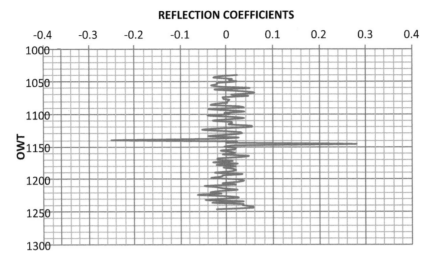

Figure 7.7 Reflection coefficients as a function of the OWT. Note the top of the gas sand at 1140 ms (the spike on the left) and the base of the gas sand at 1146 ms (the spike on the right).

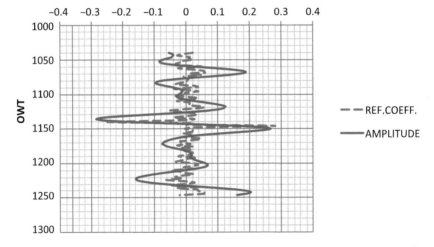

Figure 7.8 Superposition of the seismic amplitudes on the original reflection coefficients.

A common problem that is exemplified by this example is that there are other spikes in the amplitude diagram, in the middle of the shale beds at 1069, 1221 and 1240 ms. These are just artefacts of the convolution process. Although none is as strong as the actual spike, it would be easy to mistake them for evidence of extra lithological boundaries.

Even in this extremely simple example, it is clear what problems might arise from using the amplitudes as surrogates for the reflection coefficients. It is not straightforward to confirm if an anomaly is real or a convolution artefact. In general, any interpretation of seismic anomalies should be cross-checked with other sources.

The AVO method is known to be fairly reliable in practice, but should be stressed that it is a qualitative tool and that cross-checking is essential. For example, if a Class III anomaly is detected in a syncline, the result should be treated with scepticism, as it is difficult to believe there could be an accumulation of gas in such an environment.

SCALING OF THE AMPLITUDES

A seismic cube is populated by a series of seismic traces along the z or vertical direction. This vertical axis may represent depth or time, depending on how the original data were processed. The other two axes (the x and y directions) represent distances. In a 3D seismic survey, there may be a seismic trace each 15 or 20 m along the x- and y-axes.

In general, a numerical value of the seismic amplitude is registered after every 4 ms of two-way time along a seismic trace. As a first approximation, each of the amplitudes can be treated as the convolution of a string of reflection coefficients and a wavelet. If a frequency histogram is made of the amplitudes recorded in a sub-volume of the seismic cube (or over the cube as a whole), there will be a peak at amplitudes close to zero, while the remaining amplitudes are symmetrically distributed about the peak. However, the actual numerical values of the amplitudes are typically quite different from the values one might expect. An optimistic view would be that the observed amplitudes are simply proportional to the actual amplitudes, with an unknown scaling constant. For reasons that will be described in detail later, it would be helpful to know the true values of the amplitudes, or at least their statistical parameters, such as the variance or the second moment about the origin. What follows is the outline of a method for estimating the true statistical parameters of the amplitudes from the parameters of the reflection coefficients and the characteristics of the wavelet.

Consider a series of reflectivity values $x(t)$. The representation of this series in the frequency domain can be expressed symbolically as:

$$x(t) \rightarrow a - ib$$

where a and b are functions of the angular frequency.

If the amplitude series $y(t)$ corresponding to the same trace is assumed to be the convolution of a wavelet with this reflectivity series then:

$$y(t) \rightarrow (a - ib)(\alpha + i\beta) = (a\alpha - b\beta) - (a\beta + b\alpha)i$$

where α and β are functions of the angular frequency corresponding to the wavelet.

The auto-correlation of $y(t)$ in the frequency domain is therefore:

$$\phi_{yy} \rightarrow (a\alpha - b\beta) - (a\beta + b\alpha)i] [(a\alpha - b\beta) + (a\beta + b\alpha)i] = (a^2 + b^2)(\alpha^2 + \beta^2)$$

But the auto-correlations of the reflectivity series and the wavelet separately are

$$\phi_{xx} \rightarrow (a^2 + b^2)$$

and

$$\phi_{ww} \rightarrow (\alpha^2 + \beta^2)$$

so it follows that

$$\phi_{yy} = \phi_{xx} * \phi_{ww}$$

That is, in the time domain, the auto-correlation of the seismic amplitudes in a trace is the convolution between the auto-correlation of the reflectivity series and the auto-correlation of the wavelet.

However, if we consider the auto-correlation function at $\tau = 0$ (*i.e.* for zero lag), we have:

$$\phi_{yy}(0) = \phi_{xx}(0) * \phi_{ww}(0)$$

If it is assumed that the means of $x(t)$ and $y(t)$ are zero, then:

$$\phi_{yy}(0) = \sigma_y^2 N \text{ and } \phi_{xx}(0) = \sigma_x^2 N,$$

where N is the number of points in the series, and σ_y^2 and σ_x^2 are the individual variances. Then:

$$\sigma_y^2 = \phi_{ww}(0) \sigma_x^2 \tag{7.10}$$

Strictly speaking, this formula is correct only in the limit that the number of points in the series of reflection coefficients goes to infinity. Equation (7.10) demonstrates that the variance of the amplitudes is proportional to the variance of the reflection coefficients, with the constant of proportionality equal to the auto-covariate of the wavelet at zero lag. The coefficients of the wavelet are assumed to be known.

We can rewrite equation (7.10) in the form:

$$\sigma_x^2 = \sigma_y^2 / \phi_{ww}(0) \tag{7.11}$$

If the mean values of the reflection coefficients and the amplitudes are non-zero, we can use the second moment about the origin in place of the variances in equations (7.10) and (7.11). The second moment about the origin is defined by:

$$\mu_2 = \frac{\sum_{J=1}^{J=N} x_J^2}{N}$$

and is related to the mean and variance through the equation:

$$\mu_2 = \sigma^2 + \mu_1^2,$$

where σ^2 is the variance and μ_1 is the mean. When the mean is zero, the variance and the second moment about the origin coincide.

Hence, according to equation (7.11), the second moment of the reflection coefficients can be calculated if the second moment of the amplitudes and the auto-correlate of the wavelet at zero lag are known.

Equation (7.11) has been tested by simulation, using the following procedure:

1. A string of reflection coefficients was randomly generated from a normal distribution with a mean of zero and a standard deviation of 0.1 (which is equivalent to a

variance of 0.01). This effectively assumes that adjacent correlation coefficients are independent, which should be an accurate approximation to the spatial statistics of reflectivity in a real sequence. Two sets of runs were performed, with the number of reflection coefficients in the first run equal to 1000.

2. The auto-correlates of a Ricker wavelet were calculated, with a dominant frequency of 20 Hz, using 201 points.

3. The Ricker 20 wavelet was convolved with the string of reflection coefficients to generate the amplitudes.

4. The second moments of the reflection coefficients generated in this way were calculated (these are the "observed" values).

5. The predicted second moments of the reflection coefficients were calculated using equation (7.11). These are the "calculated" values. If the theory is reliable, the observed and calculated values of the reflection coefficients should be similar.

6. The process was repeated ten times to verify the consistency of the results.

7. Steps 1–6 were repeated for a second run, with a string of reflection coefficients consisting of 2000 points, to check if there was an improvement in the results.

Table 7.3 summarizes the results of this simulation.

In summary, the results of the simulation suggest that equation (7.11) is reliable. And the longer the string of reflection coefficients is, the more accurate the prediction should be.

Table 7.3 Results of the simulation exercise used to verify the reliability of equation (7.11).

Var Rc Obs	Var Rc Calc	Calc/Obs	%Error	No. RC
0.00982392	0.00944839	0.9617732	3.823	1000
0.00992664	0.00838271	0.8444659	15.553	1000
0.01053202	0.01013778	0.96256749	3.743	1000
0.0104316	0.00953491	0.91404159	8.596	1000
0.01091238	0.01426887	1.30758483	30.758	1000
0.00961034	0.00759742	0.79054647	20.945	1000
0.00968982	0.01085561	1.12031075	12.031	1000
0.0098075	0.00872183	0.88930227	11.070	1000
0.01094964	0.01242167	1.1344362	13.444	1000
		Aver Error	**13.329**	
0.00987528	0.00956694	0.96877671	3.122	2000
0.01048181	0.0096894	0.92440121	7.560	2000
0.01026136	0.01077014	1.0495821	4.958	2000
0.009774866	0.00955958	0.98060411	1.940	2000
0.01025033	0.01021173	0.99623425	0.377	2000
0.01075463	0.01068582	0.99360145	0.640	2000
0.01020516	0.00901022	0.88290869	11.709	2000
0.01008219	0.00971439	0.9635196	3.648	2000
0.00985206	0.00964528	0.97901138	2.099	2000
0.01070178	0.01080039	1.00921434	0.921	2000
		Aver Error	3.697	

The first ten runs, which yielded an average error of 13.329%, involved strings of 1000 reflection coefficients. The second set of ten runs, with an average error of 3.697%, involved strings of 2000 reflection coefficients.

THE POSSIBILITY OF ESTIMATING THE PROPORTIONS OF LITHOLOGICAL TYPES IN A RELATIVELY SMALL VOLUME

Consider now an area for which well log information of the type shown in Table 7.1 is already available. Suppose that an exploration hole is to be drilled not far from the known area, so that it is reasonable to assume that the lithological types and the fluid content will be the same as in the known area. That is, the physical properties listed in Table 7.1 are presumed to be the same in both the known area and the exploration hole, with the only difference between the two locations being the relative proportions of each lithological type. (Once again, a gas sand will be considered to be a distinct lithological type, separate from a water sand.) Under these ideal circumstances, it should be possible to estimate the proportion of each lithological type in relatively small sub-volumes of the seismic cube.

For the sake of simplicity, we will assume that just two lithological types are present: water sand and shale. Suppose we wish to estimate the proportions of sand and shale in a certain sub-volume with, say, a surface area of $10000\,m^2$ and a vertical extent equivalent to 500 ms in two-way time.

On the premise that the vertical distribution of sand and shale is completely random, the possible interfaces are sand/sand, sand/shale, shale/sand and shale/shale (where the first lithology is understood to overlie the second). Let X_{sh} be the proportion of shale in the volume and X_{sd} is the proportion of sand. It follows that $X_{sd} + X_{sh} = 1$.

The probability that a randomly chosen point in the formation is occupied by sand is X_{sd}, and the probability that the layer below is also sand is again X_{sd}. So the probability of a sand/sand interface is X_{sd}^2. Similarly, the probability of a shale/sand interface is $X_{sh} \times X_{sd}$, and so on. All four probabilities can be presented in the form of a matrix:

	SAND(down)	SHALE (dn.)
SAND (up)	Xsd^2	Xsd . Xsh
SHALE (up)	Xsh . Xsd	Xsh^2

According to this matrix, the probabilities of finding a sand/shale interface or a shale/sand interface are both $X_{sh} \times X_{sd}$. While this may be true in an infinite column of sand and shale, the proportions will never match the theoretical expectations in columns of finite extent. In a series of simulations we have performed of stratigraphic columns containing up to 2000 points, the two proportions are relatively close but never equal. Another potential source of error is that the lithological states observed in a stratigraphic column are not completely independent. If a certain point is known to be of lithology A, the probability that adjacent points will be of lithology B is not equal to the fraction of B in the entire column, but is typically a smaller conditional probability. Although lithological auto-correlation of this type can be observed in well logs, it is doubtful that it has any effects at the seismic scale. So while in principle the two elements A_{12} and A_{21} in the probability matrix could be different, they are expected in any realistic situation to be very close.

Given that $X_{sd} = 1 - X_{sh}$, it follows from the identity $(1 - X_{sh})^2 + 2X_{sh}(1 - X_{sh}) + X_{sh}^2 = 1$ that the four elements of the matrix do represent the probabilities of each of the possible types of interface in the column.

Let μ_{11} denote the second moment of the reflection coefficients of a sand/sand interface, and μ_{12}, μ_{21} and μ_{22} the analogous second moments for sand/shale, shale/sand and shale/shale interfaces, respectively. Each of these moments can be estimated through simulation, as will be discussed later. If μ_2 is the actual second moment of the reflection coefficients in the sub-volume under consideration, calculated from seismic data, then on the assumption that $\mu_{12} = \mu_{21}$ the observed second moment is in principle given by the equation:

$$\mu_2 = \mu_{11}\left(1 - X_{sh}\right)^2 + 2\mu_{12}X_{sh}\left(1 - X_{sh}\right) + \mu_{22}X_{sh}^2 \tag{7.12}$$

This equation can be solved for the unknown proportion X_{sh} to give:

$$X_{sh} = \frac{\mu_{11} - \mu_{12} \pm \sqrt{\mu_2\left(\mu_{11} + \mu_{22} - 2\mu_{12}\right) + \left(\mu_{12}^2 - \mu_{11}\mu_{22}\right)}}{\left(\mu_{11} + \mu_{22} - 2\mu_{12}\right)} \tag{7.13}$$

Before equation (7.13) can be used to calculate X_{sh}, the second moments of the reflection coefficients for the four types of interfaces considered in this example need to be estimated. It is not possible to estimate the moments analytically, so we will make use of the Monte Carlo method. All the information needed can be extracted from Table 7.1 (although in this case there are just two lithologies).

Table 7.4 shows part of a simulation exercise of this type, involving two lithologies and four types of interfaces. The steps involved in the simulation were discussed in detail earlier in this chapter. The table gives the partial results for a sand/sand interface. To estimate the second moment about the origin of the reflection coefficients for this interface (which are just the intercepts), we calculate the average of the squares of the numbers in the column "INT.(exact)". (The column to the right of this is the Aki-Richards approximation for the intercept.)

Strictly speaking, the method described above for a mixture of two lithologies does not make full use of the AVO data, because it uses only the intercept and therefore relies only on cases of normal incidence.

If there are more than two lithological types, the analysis becomes more complex. If, for example, the three lithological types present in the formation are water sand, gas sand and shale, we need to introduce an additional variable to enable us to estimate X_{sh}, X_{wsd} and X_{gsd} (the proportions of shale, water sand and gas sand, respectively). The gradient is a suitable variable for this purpose.

Given that there are three lithological components and nine possible types of interface, the interface probabilities can be represented in matrix form as follows:

	GAS SD. (dn)	WAT.SD.(dn)	SHALE (dn)
GAS SD. (up)	Xgsd ^2	Xgsd . Xwsd	Xgsd . Xsh
WAT.SD. (up)	Xwsd . Xgsd	Xwsd ^2	Xwsd.Xsh
SHALE ((up)	Xsh . Xgsd	Xsh . Xwsd	Xsh ^2

Table 7.4 Simulation of the intercept and the gradient for a sand/sand interface.

Depth	Lith_up	Lith_Down	V_{p_up}	V_{s_Down}	V_{s_up}	V_{s_Down}	R_{hob_up}	R_{hob_Down}	Int (exact)	Int.(A.R)	Gradient
510	Sand	Sand	1827.68696	1829.57254	827.719577	820.917048	2.23283515	2.2207336	-0.0021973	-0.0021972	0.00942833
511	Sand	Sand	1766.82853	1854.20148	781.465999	855.447936	1.8996897	2.10455558	0.07519865	0.07529148	-0.0915850
526	Sand	Sand	1900.37471	1724.99644	914.777235	796.07353	2.47469836	1.97169362	-0.1606227	-0.1615018	0.17600984
538	Sand	Sand	2012.59932	1909.50979	940.225013	952.292263	2.33507437	2.2034296	-0.0552483	-0.0552904	-0.0111468
540	Sand	Sand	1883.43711	1942.34677	871.823154	976.897364	2.29212465	2.1007841	-0.0281775	-0.0281586	-0.0500922
546	Sand	Sand	1783.85281	1639.84708	830.359455	721.752614	2.01477356	2.0611546	-0.0292543	-0.0292385	0.062445
549	Sand	Sand	1777.26629	1929.63899	852.296493	932.785193	2.00367469	2.01590991	0.04414348	0.044149	-0.0453672
551	Sand	Sand	1777.96875	1768.91454	859.447808	810.684439	2.05178588	1.89489677	-0.0423006	-0.0432049	0.0844923
580	Sand	Sand	1882.31075	1723.66772	960.214646	766.108661	2.08477071	2.0503602	-0.0522968	-0.052316	0.169746
589	Sand	Sand	1877.61823	1923.63245	882.5391	888.09528	2.30853538	2.24980985	9.8753E-07	9.8737E-07	0.01860932

The second moment of the intercept is given by the average of the squares of the numbers in the third-last column.

As before, the elements of the matrix represent the proportions of each type of interface in the entire column. Also, $X_{sh} + X_{gsd} + X_{wsd} = 1$, and the elements A_{ij} and A_{ji} are equal.

The system of equations analogous to (7.12) is:

$$\mu_{I2} = \mu_{I11}X_{gsd}^2 + \mu_{I22}X_{wsd}^2 + \mu_{I33}X_{sh}^2 + (\mu_{I12} + \mu_{I21})X_{gsd}X_{wsd}$$
$$+ (\mu_{I13} + \mu_{I31})X_{gsd}X_{sh} + (\mu_{I23} + \mu_{I32})X_{wsd}X_{sh} \tag{7.14a}$$

$$\mu_{G2} = \mu_{G11}X_{gsd}^2 + \mu_{G22}X_{wsd}^2 + \mu_{G33}X_{sh}^2 + (\mu_{G12} + \mu_{G21})X_{gsd}X_{wsd}$$
$$+ (\mu_{G13} + \mu_{G31})X_{gsd}X_{sh} + (\mu_{G23} + \mu_{G32})X_{wsd}X_{sh} \tag{7.14b}$$

$$1 = X_{gsd} + X_{wsd} + X_{sh}. \tag{7.14c}$$

Here, μ_{I2} and μ_{G2} are the second moments of the intercepts and gradients, respectively, which are calculated from the seismic data. Similarly, μ_{I11} and μ_{G11} are the second moments of the intercepts and gradients for a gas sand/gas sand interface, μ_{I12} and μ_{G12} are the second moments of the intercepts and gradients for a gas sand/water sand interface, and so on. If we eliminate X_{sh} by making the substitution $X_{sh} = 1 - X_{gsd} - X_{wsd}$, we are left with a system of two equations in two unknowns. Although these two equations can only be solved numerically, it is clear that even with three lithologies, it is possible to estimate the proportions of each lithology inside a certain sub-volume. For more than three lithologies, however, the problem becomes too cumbersome to attempt a solution (as another AVO parameter, such as the curvature, would need to be included in the analysis).

PROBABILITY THAT A POINT IN A GRADIENT-INTERCEPT DIAGRAM BELONGS TO AN INTERFACE OF INTEREST

Figure 7.3 plots the positions of several interfaces of interest in a gradient versus intercept diagram. The plot is based exclusively on a simulation using well data. A similar diagram could be constructed from seismic data and compared with the simulated data, but an analysis of this type is purely qualitative.

Nonetheless, the fact that the proportion of each lithology, and consequently the proportion of each type of interface, can be estimated allows us to determine the probability that any particular point in the gradient-intercept diagram belongs to a certain interface of interest. This can be done by applying Bayes' theorem, as follows.

Suppose that there are in total N types of interface, each with characteristic probability distributions for the intercept and the gradient. Let X denote the intercept, Y the gradient, and $f_k(x, y)$ the probability density function of the values x and y of the intercept and the gradient at the kth interface type. Let us assume that the probability distribution at each of the interface types is a 2-dimensional normal distribution. If a_k is the probability of the kth interface type, and at a particular point we know the values x and y of the intercept and the gradient; then according to Bayes' theorem the probability that the point belongs to the kth interface is given by:

$$P(Int.k \mid x, y) = \frac{a_k f_k(x, y)}{\displaystyle\sum_{J=1}^{J=N} a_J f_J(x, y)} \tag{7.15}$$

The following example illustrates the use of equation (7.15):

Assume there are only two possible lithological states, sand and shale, and therefore that the four available types of interface are sand/sand, shale/sand, sand/shale and shale/shale. We have already simulated a column composed only of sand and shale, part of which is shown in Table 7.4. The idea is to take all the points labelled, for example, "SAND–SAND" and calculate the means and variances of the intercept and gradient, plus the correlation coefficient. With this information, we have a complete description of the density function $f_k(x, y)$ (where "k" in this case stands for the sand/sand interface). The same procedure needs to be repeated for the other three types of interface.

If the number of interfaces in the sequence were infinite, then on the assumption that the reflection coefficients are independent, the results of the simulation would have the following properties:

1. The number of shale/sand interfaces would be the same as the number of sand/shale interfaces.
2. The means of both the intercept and the gradient would be zero for the sand/sand and shale/shale interfaces.
3. The means of both the gradient and the intercept for the shale/sand and sand/shale interfaces would have the same magnitudes but opposite signs.
4. The correlation coefficients between the intercept and the gradient would be the same for the sand/shale and shale/sand interfaces.

Table 7.5 lists the values of all the parameters needed to apply Bayes' formula (equation 7.15), calculated from a simulation involving 500 interfaces.

Although none of the properties (1–4) mentioned above is satisfied exactly, they are satisfied approximately. If 1000 or 1500 interfaces had been used, the agreement

Table 7.5 Simulated values of the parameters needed for the application of Bayes' formula.

Interface	No. of points	Fraction	Intercept		Corr. Coeff.	Gradient	
			Mean	Std. Dev		Mean	Std. Dev.
Sand/Sand	85	0.170	−0.06406	0.06406	−0.9011	0.00313	0.09557
Shale (up)/ Sand (down)	123	0.246	−0.06289	0.04937	−0.6321	−0.12197	0.07011
Sand(up)/ Shale(down)	108	0.216	0.07483	0.05480	−0.7607	0.11117	0.07426
Shale/Shale	184	0.368	−0.00343	0.04950	−0.6547	0.00525	0.0495

None of points 1–4 listed above is satisfied, because the input data is finite.

Table 7.6 Changes in the parameters from Table 7.5 made so as to adjust them to an ideal model with infinite input data.

			Mean	Std. Dev		Mean	Std. Dev
Sand/Sand	n/a	**0160**	**0.00000**	0.06406	−0.9011	**0**	0.09557
Shale (up)/ Sand (down)	n/a	**0.240**	**−0.06889**	**0.05344**	**−0.6964**	**−0.11657**	**0.7219**
Sand(up)/ Shale(down)	n/a	**0.240**	**0.06889**	**0.05344**	**−0.6964**	**0.11657**	**0.07219**
Shale/Shale	n/a	**0.360**	**0.00000**	0.04950	−0.6547	**0**	0.04950

The figures that have been changed are shown in boldface.

would probably have been ever better. Although a simulation can in principle take any number of input points, the seismic data itself are limited, and so are likely to be subject to deviations similar to those observed in the simulation. We will average the sand/shale and shale/sand parameters and make slight changes in the proportions of the interfaces so that equation (7.15) can be used.

Table 7.6 shows some slight changes in the parameters that have been made on the assumption that the proportions of sand and shale are exactly 0.4 and 0.6, respectively. Some of the numbers that have changed have been averaged, while others have been set to zero.

Suppose now that the interface of interest is shale/sand. That is, we want to estimate the probability that a certain point corresponds to an interface of that type, given the values x of the intercept and y of the gradient.

The two-dimensional normal distribution is then given by:

$$f(x,y) = \frac{1}{2\pi\sigma_X\sigma_Y\sqrt{1-\rho^2}}\exp$$
$$\times\left\{-\frac{\left[\left(\frac{x-\mu_X}{\sigma_X}\right)^2 - 2\rho\left(\frac{x-\mu_X}{\sigma_X}\right)\left(\frac{y-\mu_Y}{\sigma_Y}\right)+\left(\frac{y-\mu_Y}{\sigma_Y}\right)^2\right]}{2(1-\rho^2)}\right\} \tag{7.16}$$

The values of μ_X and μ_Y (the means of the intercept and the gradient, respectively) in this equation are taken from Table 7.6, as are the values of σ's (the standard deviations) and of ρ (the correlation coefficient).

Figure 7.9 plots the simulated data, sorted by the most likely interface type, which can be compared with a plot of the seismic data. It is clear by visual inspection in which zones of the diagram the interfaces of interest are located.

In this particular case, the type of interface we are interested in (shale overlying sand) is easy to pick visually. However, if we make use of Bayes' formula (equation 7.15), we can take any paired values of the intercept and the gradient and estimate the

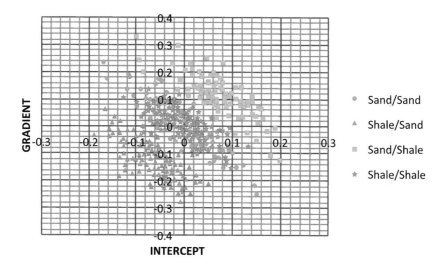

Figure 7.9 A qualitative plot, analogous to Figure 7.3, showing the regions in the gradient-intercept plane where the interfaces of interest can be found.

Table 7.7 Bayesian probabilities that points in the neighbourhood of the point

Intercept	Gradient	P(Sh-Sd|I,G)
−0.09	−0.15	0.99999987
−0.09	−0.15	0.9999996
−0.07	−0.15	0.99999883
−0.06	−0.15	0.9999965
−0.05	−0.15	0.99998944
−0.04	−0.15	0.99996728
−0.03	−0.15	0.9998944
−0.02	−0.15	0.99964111
−0.01	−0.15	0.99871578

probability that the point is located on a shale/sand interface. Table 7.7 lists those probabilities for points in the vicinity of the point (intercept = −0.05, gradient = −0.15). As would be expected from Figure 7.9, the probability that the points lie on a shale/sand interface is very high.

(intercept = −0.05, gradient = −0.15)
lie on a shalelsand interface

GENERAL COMMENTS AND SUMMARY

If we have no information about the physical properties of the rocks in an area of study, very little reliable information can be extracted from an AVO analysis. However, at shallow depths (no greater than 1500 mbsf), an interface where a shale overlies a gas sand frequently corresponds to a Class III anomaly, characterized by a negative intercept and gradient. Nonetheless, it is always advisable to cross-check with structural maps the feasibility of an accumulation of gas at the point indicated by the AVO.

A Class III anomaly is one of the four recognized classical AVO anomalies. But without the knowledge of the physical properties of the rocks in the formation, it is difficult to assess the significance of any of these anomalies (except for Class III, in the specific circumstances described above).

Even if the physical properties of the rocks are known, the AVO remains essentially a qualitative tool. A simulated plot of gradient versus intercept, generated for comparison with the intercepts and gradients calculated from the actual seismic data, is no more than a visual aid.

Any attempt to make the process of analyzing seismic data more quantitative (such as estimating the proportions of the lithological types, or the probability of a certain type of interface) depends on many assumptions, which may or may not be satisfied in practice. In addition, a quantitative algorithm will rely on good knowledge of the physical properties of the rocks in the area.

In summary, despite its limitations, the AVO technique is valuable in oil exploration, but it is important to check its results with other sources of information.

REFERENCES

Aki, K. and Richards, P.G. (1980), *Quantitative Seismology: Theory and Methods*, W.H: Freeman and Co., San Francisco.

Al-Sadi, H. (1980), *Seismic Exploration, Technique and Processing*, Birkhaüser Verlag, Basel, Boston, Stuttgart.

Rüger, A. (2002), Reflection Coefficients and Azimuthal AVO Analysis in Anisotropic Media, Geophysical Monograph Series No.10, Society of Exploration Geophysicists, Tulsa.

Sveshnikov, A., ed. (1968), *Problems in Probability Theory, Mathematical Statistics and Theory of Random Functions*, Dover Publications Inc., New York.

Zoeppritz, K. (1919), On the reflection and propagation of seismic waves, in Mavko, G. et al., eds., (1998), *The Rock Physics Handbook*, CAMBRIDGE University Press.

Chapter 8

Applications of rock physics to inversion studies

INTRODUCTION

An inversion process is, in general terms, any transformation of the seismic amplitudes (taken from a seismic section) into continuous distributions of elastic parameters, such as the P-impedance or the S-impedance. We will not discuss inversion in detail here, as it is a rather specialized field within geophysics, but instead will be concerned with the transformation of those elastic parameters into rock types (or "facies", to use a broader term). Once the lithological types in a seismic volume have been successfully identified, it may be possible to estimate the porosity and other petrophysical properties of the formation of interest, depending on the amount and quality of well information. In this chapter, we will be dealing mostly with inversion in clastic environments and will consider just one example of inversion in limestones.

An inversion is said to be "acoustic" if the only output of the process is the P-impedance. The inversion is "elastic" if it generates distributions for both the P- and S-impedances. Knowledge of these impedances allows us to calculate secondary variables, such as the quotient V_P/V_S and the products $\lambda\rho$ and $\mu\rho$. In exceptional cases, values of the density can be obtained from the gathers, by making use of the concept of "elastic impedance" introduced by Connolly (1999). Calculations of this type are limited by the quality of the gathers, but if they are possible the final products of the analysis include the density, the P-velocity and the S-velocity.

As always, it is important to combine inversion with any well information that is available for the area under study. Well data are needed to identify the rocks types present, to constrain their physical characteristics and to generate correlations between the elastic parameters and petrophysical variables, such as the porosity. Indeed, the presence of one or more wells is critical for the process of inversion itself, as realistic wavelet parameters depend on well data.

It is possible to perform an inversion without the support of well data, but the final results are more uncertain. An example will be given later of an inversion in carbonates without well control.

Also, as has been pointed out by Veeken and Da Silva (2004), the results of a seismic inversion are typically not unique. Even with access to well data, there is always some uncertainty in the values of the elastic parameters estimated from seismic data.

DOI: 10.1201/9781003261773-9

STANDARD OUTPUTS OF AN INVERSION (APART FROM THE DENSITY)

Normally, the direct products of an inversion are the two impedances, Z_P and Z_S. From these, a set of "secondary" variables can be calculated, which include

a. The ratio of the wave velocities:

$$\frac{V_P}{V_S} = \frac{Z_P}{Z_S} \tag{8.1}$$

The Poisson ratio:

$$v = \frac{\left(\dfrac{Z_p}{Z_s}\right)^2 - 2}{2\left[\left(\dfrac{Z_p}{Z_s}\right)^2 - 1\right]} \tag{8.2}$$

b. The product of the first Lamé parameter λ with the density:

$$\lambda\rho = Z_P^2 - 2Z_S^2 \tag{8.3}$$

c. The product of the second Lamé parameter μ (the shear modulus) with the density:

$$\mu\rho = Z_S^2 \tag{8.4}$$

It is evident from equations (8.1) and (8.2) that there is a functional relationship between the quotient V_P/V_S and the Poisson ratio v, so the value of one can be deduced from the knowledge of the value of the other. If we are trying to distinguish between rock types, it should be clear that there is no advantage in using both variables simultaneously. Similar remarks apply to the product $\mu\rho$ and the S-impedance, in view of equation (8.4).

So there are essentially four variables available for use in identifying rock types or "facies": (1) Z_P, (2) Z_S or $\mu\rho$, (3) V_P/V_S or v, and (4) $\lambda\rho$.

MAKING USE OF WELL DATA TO IDENTIFY FACIES AND ASSESSING THE FEASIBILITY OF AN INVERSION STUDY

In any inversion study, it is important to have access to data from at least one well, because the information provided by the well or wells allows us both to identify the facies and to determine whether it is convenient to carry out an inversion study at all.

The term "facies" is used in a loose sense here. We will define a "facies" (or an "electro-facies") to be a series of rocks – or equivalently points in a well log – that have similar physical characteristics, without reference to the sedimentary environment in which they were deposited. The terms "lithological type" or "rock type" would be synonymous to "facies" in this context.

In a clastic environment, the most fundamental distinction in rock types is between "sand" and "shale". These in turn can be sub-divided into "oil sands", "gas sands" and so on, which for present purposes are separate lithological types.

When attempting to identify the lithological types in a well, there are usually two independent sets of variables available for the analysis: the elastic parameters calculated with the Density and the two Sonic logs (and directly comparable with the seismic) and the data provided by the Gamma Ray (GR), Neutron, Resistivity and Density logs The latter are ideal for separating sand from shale, but in general are more difficult to correlate with the seismic properties of the formation. If the GR, Resistivity, Neutron and Density logs are used to identify the facies, it is clear that the elastic parameters can then be assigned to each of the points in the formation and thus independently assign elastic properties to the separate types of "sand" and "shale". The other option is to try to identify facies with the Density and Sonic logs or with elastic parameters derived from them.

It is difficult to know in advance which of the two sets of data would be best suited to the task, and some trial and error might be necessary.

There is a wide range of algorithms available for identifying facies. For detailed accounts of these, the reader is referred to the book "Statistics and Data Analysis in Geology" by Davis (1973), or the article "The portioning of petrophysical data" by Moss (1997). We will consider here only the simplest methods, involving mathematics that is not too cumbersome, which can be performed using Excel (though in some cases it might be necessary to include Visual Basic sub-routines).

USING THE PROPERTIES OF THE NORMAL DISTRIBUTION

Figure 8.1 shows a frequency histogram of GR values for a sand/shale sequence with a thickness of about 1265 m. This illustrates a situation that is relatively common in practice, where the lithological types can be identified easily using just one log. It is evident from Figure 8.1 that the log response of the sand and shale facies in this sequence could be approximated by a mixture of two normal distributions. To do this, we need to identify the statistical parameters (the mean and the standard deviation) of the distributions of the two facies, and their relative proportions. Note that, at this point, there is no reference to the elastic parameters of the facies.

The problem in this case therefore involves six unknown parameters: the two means and two standard deviations, plus the proportions of each of the two lithological components. It is possible to construct six equations for these parameters, which are mostly non-linear, although one is trivial (as the two proportions must add to one). These equations can be solved numerically using the Newton-Raphson method. The mathematical details can be found in Appendix 8.1, and the results are summarized in Table 8.1.

Once the parameters specifying the two distributions and the relative proportions of sand and shale in the sequence are known, it is possible to identify each point in the log as either "sand" or "shale", although the identification relies on Bayes' formula and is therefore probabilistic.

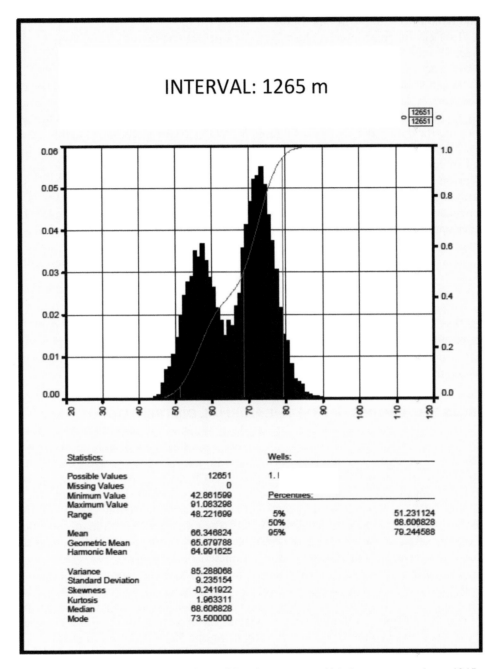

Figure 8.1 Frequency histogram of the GR values in a sand/shale sequence about 1265 m thick. The GR values are consistent with a mixture of two normal distributions, one for sand points and the other for shale points.

Table 8.1 Statistical properties of the mixture of normal distributions used to describe the GR values of the two facies.

Component	GR Mean	GR Std Dev	Proportion
Sand	58	6	0.4435
Shale	73	5	0.5565
The GR values are in API units			

An application of Bayes' formula has already been discussed in Chapter 7, and some of the misconceptions associated with Bayes' theorem are clarified in Appendix 8.2. In the case considered here, Bayes' formula for the probability that a point in the log with a GR value of x corresponds to sand is:

$$P(Sand \mid x) = \frac{\alpha_1 f_1(x)}{\alpha_1 f_1(x) + \alpha_2 f_2(x)}, \tag{8.5}$$

where α_1 is the proportion of sand in the column, α_2 is the proportion of shale (α_1 and α_2 being the two numbers in the right-hand column of Table 8.1), $f_1(x)$ is the probability density function for the GR values in sand, and $f_2(x)$ is the corresponding function for the GR values in shale. Of course, f_1 and f_2 are both normally distributed, with norm and standard deviation equal to the numbers appearing in rows 2 and 3 of Table 8.1, respectively.

If the probability calculated using equation (8.5) is greater than 0.5, the point is identified as "sand"; otherwise it is identified as "shale".

A probabilistic identification of this type is of course open to criticism. To minimize the likelihood of errors, we could instead divide the points into three classes. Those with probability greater than 0.7, say, could be treated as sand with high confidence, while those with probability less than 0.3 could be treated as shale with high confidence. However, the status of the points in between, with probabilities from 0.3 to 0.7, would then be uncertain.

Figure 8.2 plots the probability of sand given by equation (8.5) against the GR value. Two points in particular have been highlighted in blue, corresponding to the GR values at which the probability is 70% (or 0.7) and 30% (or 0.3). The corresponding critical values of GR are 63.741 API for the 70% probability and 67.065 API for the 30% probability. In other words, all GR values less than 63.741 API are to be treated as good indicators of sand, and all GR values greater than 67.065 API as good indicators of shale. Since the actual, measured distribution of the GR values is known, it is straightforward also to calculate the fraction of points that lie in the grey zone between 63.741 and 67.065 API. This fraction is a useful indicator of the "efficiency" of the measured variable (the GR value in this case) in separating two lithological types. The greater the value of this fraction, the greater will be the uncertainty in the identification of the points, and the less efficient the variable will be in separating the two rock types.

Let us consider now the situation where there are two measured variables available. In principle, the use of two measured variables should be more effective in separating

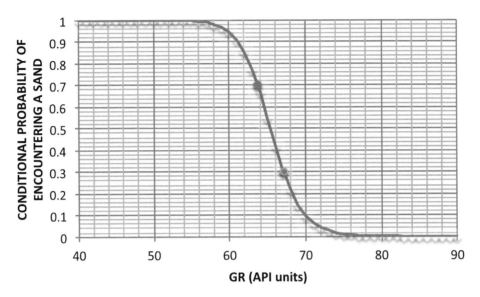

Figure 8.2 Graphical representation of equation (8.5). The two blue points correspond to the GR values at which the probability of a point being a sand are 70% and 30%, respectively.

facies than would be the case with just one. Furthermore, if the formation contains more than two facies, it would in general be very difficult to separate them using just one measured variable.

Figure 8.3 plots the points from the same 1265 m interval as shown in Figure 8.1, inside a two-dimensional parameter space. The *y*-axis in Figure 8.3 measures the GR log values, while the *x*-axis measures the values of a variable we will call *X*, which is the Neutron log porosity minus the Density log porosity. At a point where the sand is "clean", containing no shale, the value of *X* should be 0. As the proportion of shale increases, and the rock becomes more "dirty", the value of *X* should increase.

The frequency distribution in this figure is clearly bimodal. We have attempted to fit the data with a mixture of two two-dimensional normal distributions, but without success. The problem of fitting an observed distribution of points with a mixture of two two-dimensional normal distributions is simple in theory, but difficult to accomplish in practice. The theory behind a calculation of this type is described in Appendix 8.1.

It would be ideal if facies could be separated using mixtures of *n*-dimensional normal distributions in this way. But, in practice, the method works only with one or two variables, if there are no more than two possible lithological types present. Consequently, situations in which the method could be applied are rather limited.

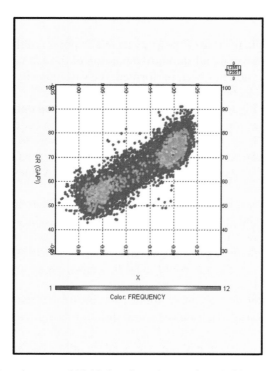

Figure 8.3 Bimodal distribution of X (defined in the text) and GR values over the same
1265 m interval as in Figure 8.1. Sands correspond to low values of X and GR.
An attempt has been made to approximate the observed values with a mixture
of two two-dimensional normal distributions.

USING CLUSTER ANALYSIS TO IDENTIFY FACIES

Suppose again that we are attempting to identify or separate the facies in a sequence,
and we have plotted a cloud of data points, such as in Figure 8.3. Intuitively, if two
points in the figure are close together, they probably belong to the same lithological
type. Conversely, if the points are widely separated in parameter space, we might infer
that they belong to different facies. Given m different measured variables, the distance
d_{ij} between two points labelled i and j in the corresponding parameter space is:

$$d_{ij} = \sqrt{\sum_{k-1}^{m}(x_{ik} - x_{jk})^2} \qquad (8.6)$$

where x_{ik} is the value of the kth variable at point i in the sequence. In order to consist-
ently weight the contributions of the m variables, these need first to be normalized,
which can be done by applying a linear transformation that shifts the mean of each
variable to 0 and rescales each standard deviation to 1. The explicit formula for the
normalized values x_{jk} is:

$$x_{jk} = \frac{y_{jk} - \mu_k}{\sigma_k}, \tag{8.7}$$

where y_{jk} is the original value of the kth variable at point i in the sequence, and μ_k and σ_k are the measured mean and standard deviation of the kth variable.

Once normalization has been performed, the vast majority of the values of each of the m variables will lie between −3 and 3.

It is clear from Figure 8.3 (which plots non-normalized data) why normalization is needed. The numbers on the x-axis are an order of magnitude smaller than the numbers on the y-axis. If the distances between the points were to be calculated using the original values of the variables, the differences between the GR values would dominate, while the differences between the much smaller X values would have almost no effect on the distances.

Equation (8.6) has been taken from Davis (1973). There are many different clustering methods available in the geological literature, and the simple method based on equation (8.6) that we will use here has no particular name. The method pre-supposes that some basic geological knowledge is available about the area from which the data set is taken. For example, we would need to know how many lithological types are represented in the data.

We will consider now a simple application of the method. In this example, there are two measured variables, X_1 and X_2, and the data set consists of seven points, which are listed in Table 8.2, and plotted in Figure 8.4. It should be noted that, in this particular example, no normalization of the original data points has been performed, because both variables are of the same order of magnitude. The question of interest here is whether the observed points can be separated into distinct groups.

Table 8.3 shows the mutual distances between all the points. Two points, i and j, are said to be "nearest neighbours" if i is the closest point to j, and j is the closest point to i. The highlighted cells in Table 8.3 indicate which pairs of points are nearest neighbours.

Consider, for example, the first row of Table 8.3, which lists the distances between point 1 and the other six points. It can be seen that the closest point to point 1 is point 2, at a distance of 3.6. However, from the second row of the table, it is evident that the closest point to point 2 is point 3 (at a distance of 2), not point 1. So points 1 and 2 are not nearest neighbours, and we move on to consider point 2. We know already that point 3 is the closest point to point 2, and it can be seen from row three that point 2 is

Table 8.2 Values of the variables X_1 and X_2 at the seven points

X_1	X_2	Pt. Label
3	4	1
5	7	2
7	7	3
12	8	4
15	11	5
17	12	6
18	13	7

Table 8.3 Mutual distances between the points in Table 8.2.

Points	1	2	3	4	5	6	7
1	-	3.6	5	9.8	13.9	16.1	17.5
2	3.6	-	2	7.1	10.8	13	14.3
3	5	2	-	5.1	8.9	11.2	12.5
4	9.8	7.1	5.1	-	4.2	6.5	7.8
5	13.9	10.8	8.9	4.2	-	2.2	3.6
6	16.1	13	11.2	6.5	2.2	-	1.4
7	17.5	14.3	12.5	7.8	3.6	1.4	-

The highlighted cells indicate which pairs of points are "nearest neighbours".

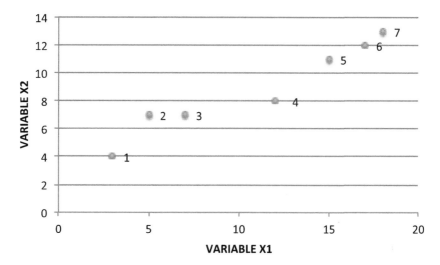

Figure 8.4 Distribution of points in the parameter space.

the closest point to point 3. So points 2 and 3 are nearest neighbours. The only other pair of nearest neighbours in Table 8.3 is points 6 and 7, at a distance of 1.4.

Once all the pairs of nearest neighbours have been identified, they are replaced in the mutual-distance table by their "averages", which in this case are the points midway between them. In later rounds, the "averages" are weighted by the number of original points n_j and n_k that are already represented by the points j and k being averaged. In general, the two formulas for the coordinates (X_{1NEW}, X_{2NEW}) of each averaged point are:

$$X_{1NEW} = \frac{n_j X_{1j} + n_k X_{1k}}{n_j + n_k}$$

$$X_{2NEW} = \frac{n_j X_{2j} + n_k X_{2k}}{n_j + n_k}$$

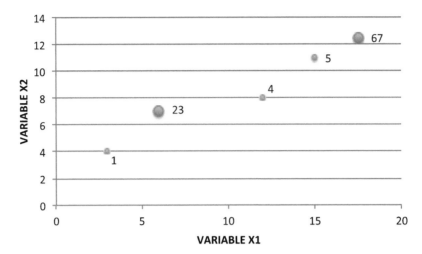

Figure 8.5 Distribution of points after replacing points 2 and 3 with point 23 and points 6 and 7 with point 67.

where (X_{1j}, X_{2j}) and (X_{1k}, X_{2k}) are the coordinates of the points j and k just before the averaging is performed. In the first round of averaging, the values of n_j and n_k are always 1. However, they may be greater than 1 in later rounds.

To continue the example, the first round of averaging involves replacing points 2 and 3 with their midpoint, which is point 23 in Figure 8.5, and points 6 and 7 with their midpoint, which is point 67 in Figure 8.5.

Table 8.4 lists the mutual distances between the 5 points remaining after the first round of averaging.

It is evident from Table 8.4 that points 1 and 23 are now nearest neighbours, as are points 5 and 67. In the second round of averaging points 1 and 23 are therefore replaced with point 123, and points 5 and 67 with point 567. It is important to be aware that the coordinates of point 123 are calculated with $n_1 = 1$ and $n_{23} = 2$ and similarly that the coordinates of point 567 are calculated with $n_5 = 1$ and $n_{67} = 2$.

After the second round of averaging, the data set has been reduced to just three points. It is straightforward to verify at this stage that points 4 and 567 are now nearest neighbours, so at the end of the process there are two aggregate points, 123 and 4567.

Table 8.4 Mutual distances between the points once points 2 and 3 have replaced with point 23 and points 6 and 7 with point 67.

Points	1	23	4	5	67
1	-	4.2	9.8	13.9	16.8
23	4.2	-	6.1	9.8	12.7
4	9.8	6.1	-	4.2	7.1
5	13.9	9.8	4.2	-	2.9
67	16.8	12.7	7.1	2.9	-

The highlighted cells indicate the new pairs of nearest neighbours.

These could in principle be identified as representatives of two different rock types, one comprising 3/7 of the data points, and the other the remaining 4/7.

Each point is therefore assigned to the rock type whose centre of gravity is closer to the point in question. Note that there is a certain degree of subjectivity involved in the clustering process. We could have stopped at points 123, 4 and 567, leaving three facies comprising 3/7, 1/7 and 3/7 of the data points, or we could have continued by combining all seven points into one facies. It is for this reason that independent geological and petrophysical information is needed to constrain the number of facies.

The method described above is simple to understand and is relatively easy to implement using Visual Basic or an equivalent programming language. The method is also quite general, in the sense that any set of variables could in principle form a basis for identifying the facies. In practice, however, facies are normally identified from the values of the elastic variables measured through seismic inversion.

Of course, many alternative options are available. One possibility is to identify the facies using the GR, Resistivity, Neutron and Density logs (which are excellent tools for discriminating sand from shale), then construct facies-specific distributions of the elastic variables (the P- and S-impedances, or parameters derived from them). These distributions could in turn be used to refine the assignment of points to individual facies, although this can be a laborious trial-and-error process if done properly. As an example, the facies in Figure 8.6 were originally identified using the X and GR values (see the third track from the right), then the classification of the individual points in the column was refined by propagating with the elastic variable V_p/V_s (see the second track from the right).

What is meant here by "propagating" with an elastic variable perhaps needs some clarification. The two facies, sand and shale, are first identified using the nearest neighbour method, with X and GR as the cluster variables. Choosing any set of elastic variables, we then calculate the centres of gravity (the means) and the standard deviations corresponding to the two groups of data points. If two or more elastic variables are involved in the analysis, their correlation coefficients are calculated as well. (The standard deviations and correlation coefficients are intended for later applications.) Each point in the column is then reassigned to the facies whose centre of gravity it is closer to.

Note that, in this case, a single elastic variable is sufficient to discriminate sand from shale. The method would be extremely valuable if it could be extended to discriminate between the principal types of sands, namely, water sands, oil sands and gas sands. The initial classification is made using well log data alone, and it would need to be checked if any combination of elastic variables could refine the classification so as to reliably distinguish between the various types of sands – and in particular, to separate water sands from hydrocarbon-bearing sands.

For each facies or lithological type, the distributions of the elastic variables will not necessarily be separately normal, and in principle any unimodal distribution could underlie each variable. However, in the analysis that follows we will attempt to apply Bayes' theorem to the elastic variables, and the calculations are easier if one assumes that the underlying distributions are normal. For a combination of n variables, Bayes' formula can be written in the form:

$$P(Facies_k \mid x_1, x_2, ... x_n) = \frac{\alpha_k f_k (x_1, x_2, ... x_n)}{\displaystyle\sum_{j=1}^{j=m} \alpha_j f_j (x_1, x_2, ... x_n)} \tag{8.8}$$

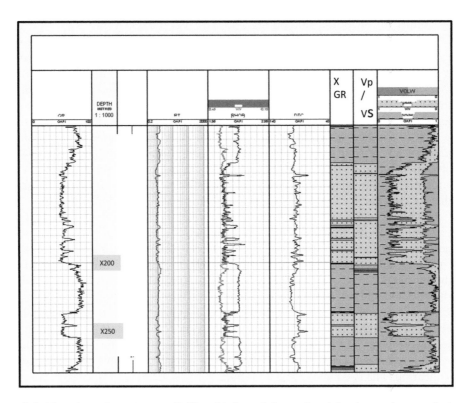

Figure 8.6 Identifying facies in a well. The third track from the right shows the two facies identified using the variables X and GR. The second track from the right shows the facies identified again from the variables X and GR, but propagated with the elastic parameter V_p/V_s. The rightmost track shows the well log evaluations, from which the effective porosity and the volume fraction of shale (V_{sh}) have been calculated.

where $f_k(x_1, x_2, \ldots, x_n)$ denotes the n-dimensional normal distribution, evaluated at particular values $x_1, x_2, \ldots, x_n$ of the variables in the kth facies.

SCALING THE WELL DATA TO MAKE IT COMPATIBLE WITH THE SEISMIC DATA

Up to this point, it has been assumed that the initial identification of the facies is made exclusively with well data and that the elastic variables are then introduced to cross-check and possibly refine the classification. Well data have a resolution of the order of centimetres, whereas the seismic resolution is of the order of metres. The results of a seismic inversion are of little value if they cannot discriminate facies at both resolutions. In order to check if the well data are consistent with facies discrimination at the seismic level, it is necessary to "upscale" the well data to seismic resolution. The standard technique for doing this is known as Sequential Backus Averaging and is described, for example, by Lindsay and Van Koughnet (2001).

Backus averaging involves calculating the harmonic mean of the bulk and shear moduli and the arithmetic mean of the bulk density over an averaging window whose width has to be determined. It is assumed that the beds are horizontal.

As a rule of thumb, the width of the averaging operator applied to the well data is taken to be $\lambda/4$, where λ is a characteristic wavelength of the seismic waves. In any seismic sub-volume of interest, it is always possible to identify the dominant frequency f of the seismic waves. The average velocity v of the waves inside this sub-volume can also be calculated from the seismic velocities, so $\lambda = v/f$ is easily estimated. For example, if the dominant frequency of the seismic waves is $f = 20\,Hz$ and the average velocity is $v = 3000\,m/s$, then $\lambda = 150\,m$ and $\lambda/4 = 37.5\,m$. The last number is the width of the averaging window. In a well log, a sample is typically taken every $0.1524\,m$, so the number of points in the well log that have to be averaged in each step is about 247. For optimum results, perhaps the best way to establish the width of the smoothing operator is by trial and error, with the values of parameters derived from the seismic data compared with the corresponding values calculated from the averaged well log data using different widths of the smoothing operator.

Figure 8.7 plots the original bulk modulus, calculated from a set of well data, and the smoothed bulk module at what should be seismic resolution, as functions of depth.

It should be emphasized that only the bulk and shear moduli and the density are calculated by smoothing in the Backus method. The averages of the bulk and shear moduli are their harmonic means, while the average of the density is its arithmetic mean. The values of all other elastic parameters at seismic resolution are calculated directly from these three smoothed parameters.

An examination of Figure 8.7 suggests that the averages and standard deviations of the elastic variables for each of the lithological types will need to be reviewed when the well data are compared directly with the seismic data. Over the interval of roughly 600 m shown in Figure 8.7, the mean of the bulk modulus taken from the original well

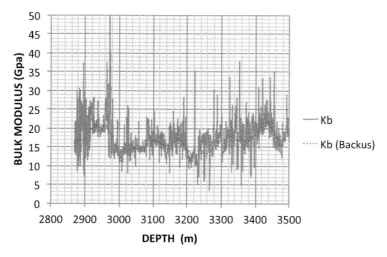

Figure 8.7 The original bulk modulus, at well resolution, and the smoothed bulk modulus, at seismic resolution, as functions of depth.

data is 17.46 GPa, whereas the mean of the smoothed bulk modulus is 16.45 GPa. Similarly, the standard deviations are 4.567 and 2.340 GPa, respectively. These differences are substantial, although it is to be expected in general that the smoothed data will have a significantly smaller standard deviation than the original data. The smoothing process should, of course, be repeated for the shear modulus and the density, and any other elastic parameters of interest calculated from the smoothed values. At this point, the smoothed elastic parameters constructed from the well data should be at the same scale as the elastic variables generated by the seismic inversion.

A QUICK RECAPITULATION OF THE STEPS INVOLVED IN ANALYSING WELL DATA

1. Identifying the facies. The analyst must decide how many lithological types or facies are present in the well data. The facies may be identified using the well log variables (taken from the Neutron, GR and Bulk Density logs), then refined by propagating with the elastic variables (again taken from the well log data). Alternatively, the facies can be discriminated directly using the elastic variables. In either case, the facies can be identified using the nearest neighbour analysis described above. That is, each point in the log is assigned to the closest facies, based

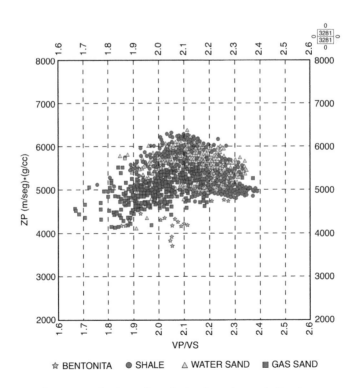

Figure 8.8 Stratigraphic interval where four facies have been defined.

Table 8.5 Mean parameters of the four facies, calculated from the well data, plus the number of points NP represented by the facies in the column, and the relative proportion α of the facies in the column.

| | V_p/V_s | | Z_p | | | | |
	μ	σ	μ	σ	r	NP	α
Bentonite	2.0340	0.1086	4444.3	365.48	0.1860	29	0.0088
Shale	2.1694	0.0812	5501.8	363.44	0.4876	2418	0.7370
Water sand	2.1442	0.1014	5545.4	328.54	0.2446	421	0.1283
Gas sand	1.9982	0.1081	5004.1	332.93	0.5558	423	0.1259

Once the mean values have been reviewed after smoothing, they will be used to populate the seismic cube with facies.

on the distances (in the multidimensional normalized parameter space) between the centres of gravity of all the facies and the point in question. At this stage, after constructing crossplots and performing a lot of trial-and-error work, the analysist should know if an inversion study is feasible.

2.   If the analyst decides, in Step 1, that an inversion is feasible, the values of all the elastic variables that were used to discriminate the facies should be smoothed to make them compatible with the seismic data. The mean parameters estimated for each facies will of course change after smoothing.

Figure 8.8 and Table 8.5 present an example of the first step described above. Figure 8.8 is a crossplot from which four facies have been identified: bentonite, water sand, gas sand and shale. The data that are plotted here are unsmoothed well data. Table 8.5 lists the mean parameters for all the facies, prior to smoothing. When it comes to populating the seismic cube with facies, the number of facies will be reduced to three, as the bentonite beds are too thin to be included.

Incidentally, it is evident from Figure 8.5 that the separation of the water sands and the shale is problematic, to say the least. However, the gas sands are clearly separated from the water sands and the shale, so an inversion study is still worthwhile.

POPULATING THE SEISMIC CUBE WITH FACIES

Once the geophysical task of populating several seismic cubes with the estimated values of the elastic variables has been completed, the job of the analyst is assign each of the points in the cubes to one of the facies identified from the well data. We will concentrate here on two basic methods for doing this. The advantage of these methods is that there are no "black boxes" involved, so the analyst knows exactly what is being calculated at each step. It should be pointed out that, when classifying the points in a seismic cube in this way, the only thing that changes is the proportions of the facies in each vertical section. The mean values of the parameters for the different facies remain unchanged over the whole cube.

Of the two methods that will be discussed here, the first assumes that the relative proportions of the facies are known, or can be estimated, while the second is intended for situations where the relative proportions are unknown.

1. If the relative proportions of the facies are known: use Bayes' formula
    This would be the ideal method, because not only would it assign a facies to each point but it would also generate the set of probabilities that the point belongs to any one of facies. However, this method is difficult to apply in practice, as it requires reliable estimates of the proportion of each facies in the sub-volume under investigation.

It is clear that the relative proportions α_k of the facies are critical inputs to Bayes' formula (equation 8.7). Assume by way illustration that there are three facies in the sub-volume and that two variables are used to discriminate between them.

Let μ_1 and μ_2 be the overall means of variables 1 and 2 in the sub-volume, μ_{11}, μ_{12} and μ_{13} the means of variable 1, and μ_{21}, μ_{22} and μ_{23} the means of variable 2 for facies 1, 2 and 3, respectively, and α_1, α_2 and α_3 the proportions of the three facies.

These numbers will satisfy the following set of equations:

$$\mu_1 = \mu_{11}\alpha_1 + \mu_{12}\alpha_2 + \mu_{13}\alpha_3 \tag{8.9a}$$

$$\mu_2 = \mu_{21}\alpha_1 + \mu_{22}\alpha_2 + \mu_{23}\alpha_3 \tag{8.9b}$$

$$1 = \alpha_1 + \alpha_2 + \alpha_3 \tag{8.9c}$$

which constitute a system of three linear equations for the three unknowns α_1, α_2 and α_3. The solution is straightforward and reads:

$$\alpha_2 = \frac{\dfrac{\mu_1 - \mu_{13}}{\mu_{11} - \mu_{13}} - \dfrac{\mu_2 - \mu_{23}}{\mu_{21} - \mu_{23}}}{\dfrac{\mu_{12} - \mu_{13}}{\mu_{11} - \mu_{13}} - \dfrac{\mu_{22} - \mu_{23}}{\mu_{21} - \mu_{23}}} \tag{8.10}$$

and

$$\alpha_1 = \frac{(\mu_2 - \mu_{23}) - \alpha_2(\mu_{22} - \mu_{23})}{\mu_{21} - \mu_{23}} \tag{8.11}$$

(while of course $\alpha_3 = 1 - \alpha_1 - \alpha_2$). Equations (8.10) and (8.11) are potentially very useful because problems involving three facies and two variables are fairly common. Unfortunately, a major drawback is that quite often the values of the proportions calculated in this way do not lie in the physical range between 0 and 1. This inconsistency can be attributed partly to the fact that the properties of each lithological type are normally not constant across the cube, and so are only crudely represented by their mean values, and partly to the fact that the model itself is imperfect. As a result, small errors in the input parameters can create serious problems when estimating the proportions.

A necessary, but not sufficient, condition for equations (8.9a)–(8.9c) to have a solution that is physically viable is for the overall mean value μ_j of each variable to lie between the corresponding means μ_{j1}, μ_{j2} and μ_{j3} for the three facies. For instance, if the values of V_p/V_s for the three facies are 1.95, 2.04 and 2.16, then the mean value of V_p/V_s should lie between 1.95 and 2.16. If it does not, it is clear that the three proportions cannot lie in the range of 0–1. This is a common problem, but it does not necessarily mean that Bayes' formula should be abandoned, as other methods are available for estimating the proportions, as will be seen shortly.

It should be mentioned that the same problem exists in the simpler case of two facies and one variable, where the relation between the means is $\mu_1 = \alpha_1 \mu_{11} + (1 - \alpha_1) \mu_{12}$, which solves to give:

$$\alpha_1 = \frac{\mu_1 - \mu_{12}}{\mu_{11} - \mu_{12}}. \tag{8.12}$$

Again, the value of α_1 will lie in the physical range of 0–1 only if μ_1 lies between μ_{11} and μ_{12}.

Consider now the situation where there are three facies and three variables, which is also a rather common case. The system of equations analogous to (8.9a)–(8.9c) is now overdetermined, as there are more equations (*i.e.* equation 8.4) than there are unknowns equation (8.3). It is always possible in this case to ignore one of the variables and attempt to solve for the proportions using the other two variables alone. If it is not possible to generate estimates of the proportions in the range 0–1 in this way, an alternative strategy is to try to solve the overdetermined system of three variables, while ignoring the constraint $\alpha_1 + \alpha_2 + \alpha_3 = 1$. There is then no guarantee that the proportions of the three facies will add to one, but if the sum is close to one (as a rough guide, if it lies between 0.95 and 1.05) and none of the calculated fractions is negative, the proportions can be accepted, then rescaled to exact proportions by dividing each one by the sum.

Yet another possibility is to use the method of ordinary least squares to find an approximate solution. The overdetermined system analogous to equations (8.9a)–(8.9c) is:

$$\mu_1 = \mu_{11}\alpha_1 + \mu_{12}\alpha_2 + \mu_{13}\alpha_3$$
$$\mu_2 = \mu_{21}\alpha_1 + \mu_{22}\alpha_2 + \mu_{23}\alpha_3$$
$$\mu_3 = \mu_{31}\alpha_1 + \mu_{32}\alpha_2 + \mu_{33}\alpha_3$$
$$1 = \alpha_1 + \alpha_2 + \alpha_3$$

Again, the unknown parameters are the three α's. If we now define the residuals R_j of the four equations to be:

$$R_1 = \mu_{11}\alpha_1 + \mu_{12}\alpha_2 + \mu_{13}\alpha_3 - \mu_1$$
$$R_2 = \mu_{21}\alpha_1 + \mu_{22}\alpha_2 + \mu_{23}\alpha_3 - \mu_2$$
$$R_3 = \mu_{31}\alpha_1 + \mu_{32}\alpha_2 + \mu_{33}\alpha_3 - -\mu_3$$
$$R_4 = \alpha_1 + \alpha_2 + \alpha_3 - 1$$

then the objective of the method of ordinary least squares is to choose the α's so that the sum of the squares of the residuals is as small as possible. This can be done by setting the three derivatives of the function $R_1^2 + R_2^2 + R_3^2 + R_4^2$ with respect to α_1, α_2 and α_3 equal to 0. This results in a system of three equations in the three unknowns, which is readily solved. Of course, since the solution typically has $R_4 \neq 0$, the three proportions will not in general add to one. It is therefore up to the analyst to decide whether the proportions estimated in this way have any practical value.

Once the proportions of all the facies or lithological types in the sub-volume of interest are known, Bayes' formula (equation 8.8) can be applied to calculate the probability that each point in the sub-volume belongs to a particular facies. Referring back to equation (8.8), it has already been mentioned that the joint probability density function $f_k(x_1, x_2, ..., x_n)$ of the n variables in the kth facies is usually assumed to be normal. Of course, there is no guarantee that the joint distribution of the variables on a set of points constructed using the nearest neighbour method will in reality be close to normal, but this presumption is a fundamental feature of the model.

The joint probability density function for the n-dimensional normal distribution is (Sveshnikov, 1968):

$$f(x_1, x_2, ..., x_n) = \frac{1}{(2\pi)^{n/2}\sqrt{\Delta}} \exp\left\{-\frac{1}{2}\left[\sum_{i=1}^{n}\sum_{j=1}^{n}\frac{A_{ij}}{\Delta}(x_i - \mu_i)(x_j - \mu_j)\right]\right\} \tag{8.13}$$

where Δ is the determinant of the matrix of variances and covariances, and A_{ij} is the cofactor of the element k_{ij} in the same matrix.

As an illustrative example, consider the case of $n = 2$ variables. The matrix of variances and covariances is then:

$$\begin{pmatrix} k_{11} & k_{12} \\ k_{21} & k_{22} \end{pmatrix} = \begin{pmatrix} \sigma_1^2 & \rho\sigma_1\sigma_2 \\ \rho\sigma_1\sigma_2 & \sigma_2^2 \end{pmatrix} \tag{8.14}$$

That is to say, it is a symmetric matrix and whose diagonal entries are the variances of the two variables and whose off-diagonal entries are the covariance. (Here, ρ is the correlation coefficient between the variables.) The determinant of the matrix (equation 8.14) is:

$$\Delta = \sigma_1^2\sigma_2^2 - \rho^2\sigma_1^2\sigma_2^2 = \sigma_1^2\sigma_2^2(1 - \rho^2) \tag{8.15}$$

The cofactor of the element k_{ij} is the determinant of the reduced matrix generated by eliminating row i and column j from the original matrix. In the case considered here, the reduced matrix consists of just a single number, which is the cofactor itself. For example, if $i = 1$ and $j = 1$ then the first row and first column of equation (8.14) are eliminated, leaving the element $k_{22} = \sigma_2^2$, which is the cofactor (i.e. $A_{11} = \sigma_2^2$).

Consider now the double sum in the density function (equation 8.13). This contains $n = 4$ terms, which are:

(when $i = 1$ and $j = 1$, so that $A_{11} = \sigma_2^2$) $\sigma_2^2 (x_1 - \mu_1)^2$ (8.16a)
(when $i = 1$ and $j = 2$, so that $A_{12} = \rho \sigma_1 \sigma_2$) $\rho \sigma_1 \sigma_2 (x_1 - \mu_1)(x_2 - \mu_2)$ (8.16b)
(when $i = 2$ and $j = 1$, so that $A_{21} = \rho \sigma_1 \sigma_2$) $\rho \sigma_1 \sigma_2 (x_1 - \mu_1)(x_2 - \mu_2)$ (8.16c)
(when $i = 2$ and $j = 2$, so that $A_{22} = \sigma_1^2$) $\sigma_1^2 (x_2 - \mu_2)^2$ (8.16d)

After adding the expressions (8.16a)–(8.16d) and then dividing the sum by the determinant (8.15), the argument of the exponential function in equation (8.13) becomes:

$$ARG = -\frac{1}{2(1-\rho^2)}\left[\frac{(x_1 - \mu_1)^2}{\sigma_1^2} + \frac{2\rho(x_1 - \mu_1)(x_2 - \mu_2)}{\sigma_1\sigma_2} + \frac{(x_2 - \mu_2)^2}{\sigma_2^2}\right]$$

and equation (8.13) reads:

$$f(x_1,x_2) = \frac{1}{2\pi\sigma_1\sigma_2\sqrt{1-\rho^2}} \exp\left\{-\frac{1}{2(1-\rho^2)}\left[\begin{array}{l}\dfrac{(x_1-\mu_1)^2}{\sigma_1^2} + \dfrac{2\rho(x_1-\mu_1)(x_2-\mu_2)}{\sigma_1\sigma_2} + \\[2mm] \dfrac{(x_2-\mu_2)}{\sigma_2^2} + \dfrac{(x_2-\mu_2)^2}{\sigma_2^2}\end{array}\right]\right\} \quad (8.17)$$

A distribution function of this type can in principle be constructed for each of the facies, using the means, variances and correlation coefficients calculated from the well data. So the joint distributions of the variables in each facies can always be expressed, in the general case, in terms of n-dimensional normal distributions. These distributions are then combined in equation (8.8) with the relative proportions of the facies in the cube to give the probability that a particular point with known values of the variables belongs to facies k.

2.   If the relative proportions of the facies are unknown: use nearest neighbour analysis.

If Bayes' formula (equation 8.8) cannot be applied, it is still possible to assign the points in the seismic cube to the facies using nearest neighbour analysis, although in this case there is no reliable way of estimating the probability that any point really belongs to the facies it has been assigned to.

As an example, consider a case involving two variables and three facies, whose centres of gravity in the normalized two-dimensional space are shown in Figure 8.9.

The procedure that needs to be followed when normalizing the variables was explained earlier in this chapter, and in particular the coordinates of the normalized data points are given by equation (8.7).

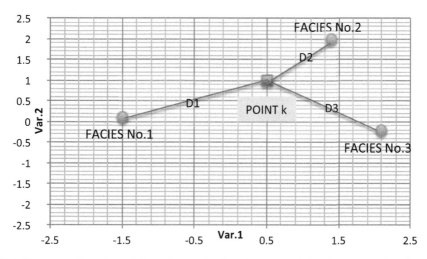

Figure 8.9 Centres of gravity of three facies (defined using well data) in normalized coordinates. Point k in the seismic volume will be assigned to facies no. 2, because this is the closest of the three facies.

Once the elastic variables across the seismic volume have been normalized, the "distance" between each point in the volume (such as the point k in Figure 8.9) and facies i is calculated using a formula analogous to equation (8.6). In the case shown in Figure 8.9, for example, the distance D_i from the unassigned point to facies i is given by

$$D_j = \sqrt{\sum_{k=1}^{2} \left(x_k - \mu_{jk}\right)^2} \tag{8.18}$$

where (x_1, x_2) are the coordinates of the unassigned point, and (μ_{i1}, μ_{i2}) are the coordinates of the centre of gravity of facies i.

In the general case, we calculate as many distances as there are facies present. The point in question is then assigned to the facies with the smallest value of the distance D_j. In this fashion, the entire seismic cube can be populated. Note, however, that Bayes' formula approach described earlier is preferable, because it gives additional information about how certain we are that any point belongs to a particular facies. Only if Bayes' formula cannot be applied should the nearest neighbour approach be used.

ESTIMATION OF THE EFFECTIVE POROSITY OF A "SAND" FACIES

In order to estimate the effective porosity of a sand, well data are needed, so that a correlation can be constructed between the effective porosity in the well and one or more of the elastic variables. As a first step, any correlation should be constructed using the original variables at well log scale, then translated into smoothed variables, to make it compatible with the seismic data.

In principle, there should be a tight relationship between the effective porosity and the P-impedance in sands. If the Wyllie equation is assumed to be approximately valid for sands, it follows that:

$$Z_p = V_p \rho_b = \frac{304.8[\rho_{ma} - (\rho_{ma} - \rho_f)\phi]}{\Delta T_{ma} + (\Delta T_f - \Delta T_{ma})\phi} \tag{8.19}$$

Here, the denominator is assumed to be measured in μs/ft, and the factor 304.8 is needed to convert the impedance to (km/s)(g/cc). Because the transit time ΔT_{ma} and density ρ_{ma} in the matrix are typically not constant – they change as the proportion of shale changes – and because the Wyllie equation is only approximately valid, a crossplot of the effective porosity against the P-impedance will produce a cloud of points rather than a line. Figure 8.10 shows a crossplot of this type, with the points corresponding to unsmoothed data taken from a sand facies. Note that although the relation between Z_P and ϕ predicted by equation (8.19) is non-linear, the data in Figure 8.10 very closely follow a straight line.

In normal circumstances, four independent elastic variables should be available $(Z_p, Z_s, Z_p/Z_s$ and $\lambda\rho)$, and it is worthwhile checking if any improvement in the predicted values of the porosity will result if several of these variables are combined. To do this, we can perform a least-squares minimization, assuming that a functional relationship exists between the predicted variable y (which in this case would be the porosity) and the input variables $x_1, x_2, ..., x_n$. Let y be the variable to be calculated and x_1, $x_2, ..., x_n$ the input variables. The Standard Error of the Estimate is then defined to be:

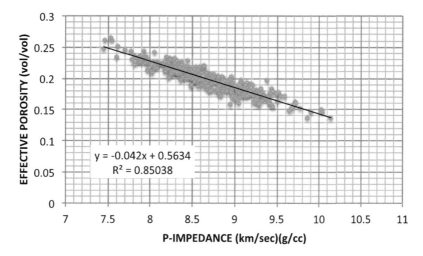

Figure 8.10 Crossplot of effective porosity vs. P-impedance in a sand facies, using un-smoothed data. The relation between the two variables is closely approxi-mated by a straight line.

$$S_y = \sqrt{\frac{1}{N}\sum(y - y_{est})^2} \tag{8.20}$$

where N is the total number of data points, the y are the observed values of the variable to be predicted (the porosity) and the y_{est} are the estimated values of the same variable. The least-squares fit to the data involves choosing a functional combination of the input variables that minimizes the value of s_y. Correlations with other variables, such as the permeability, can also be attempted.

A THEORETICAL EXAMPLE INVOLVING INVERSION IN CARBONATES

We will consider now a case in which it is assumed that the bulk density has been extracted from results of the inversion process, which is not often possible. The rock mass to be analysed in this example contains just dolomite and calcite, plus of course pore space, so "shale" and "clay" are completely absent. There is also no gas in the formation, and we will not be able to discriminate between oil and water. Under these very specialized conditions, knowledge of the density and the P- and S-velocities allows us to calculate the porosity and the volume fractions of calcite and dolomite at every point within the seismic volume. This can be done by applying the principles of rock physics alone, as it is assumed that there are no wells nearby from which correlations could be constructed.

Since we know the density ρ and the seismic velocities V_p and V_s at each point, we can easily calculate the corresponding values of the bulk modulus K_b and the shear modulus μ. Using these values as starting points, we will calculate the effective porosity ϕ (which is identical to the total porosity because there is no "shale" present) and the volume fractions of calcite and dolomite.

In any elastic material which is itself a mixture of other materials, the effective value M of any particular elastic parameter can be written as:

$$M = wM_{SUP} + (1-w)M_{INF}, \tag{8.21}$$

where M_{SUP} and M_{INF} are the maximum and minimum possible values of the parameter, given by the Hashin-Shtrikman upper and lower bounds, respectively (see Chapter 5), and w is a "weight" ranging between 0 and 1.

For any rock that can be treated as a mixture of "matrix" and "fluid" components, the upper or maximum and lower or minimum bounds of the bulk modulus are:

$$K_{b\,max} = K_0 + \frac{\phi}{(K_F - K_0)^{-1} + (1-\phi)\left(K_0 \frac{4}{3}\mu_0\right)^{-1}} \tag{8.22}$$

and

$$K_{b\,min} = \frac{K_F K_0}{K_F(1-\phi) + K_0\phi}, \tag{8.23}$$

where K_F is the incompressibility of the fluid component and K_0 is the incompressibility of the solid part of the rock (the matrix). Because the rock is a mixture of calcite and dolomite in unknown proportions, K_0 at this stage is unknown. We will assume that the fluid is water, as not enough information is available with seismic data alone to discriminate between oil and water. From equation (8.21), it follows that:

$$K_b = wK_{b\,max} + (1-w)K_{b\,min} \tag{8.24}$$

At this stage we introduce the Marion hypothesis (*cf.* Chapter 5), which assumes that the weight w appearing in equation (8.24) is independent of the type of fluid occupying the pore space. We can push this hypothesis to the limit by assuming that it applies also to the dry rock, so that

$$K_{dry} = wK_{dry\,max} + (1-w)K_{dry\,min}, \tag{8.25}$$

where the weights w in equations (8.24) and (8.25) are the same.

The dry rock is obviously devoid of fluid, so $K_F = 0$ in equation (8.23) and the minimum possible value of K_{dry} is 0. Similarly, from equation (8.22), the maximum possible value of K_{dry} is:

$$K_{dry\,max} = K_0 + \frac{\phi}{(1-\phi)\left(K_0 \frac{4}{3}\mu_0\right)^{-1} - K_0^{-1}} \tag{8.26}$$

In addition, we can make use of the Krief et al. correlation from Chapter 6:

$$\frac{K_{dry}}{\mu} = \frac{K_0}{\mu_0}, \tag{8.27}$$

which states that the ratio of the incompressibility to the shear modulus in the dry rock is identical to the same ratio in the solid part of the rock.

The porosity ϕ can be expressed as a function of the densities ρ_{ma}, ρ_b and ρ_f by solving equation (8.12) in Chapter 1 (with $V_{sh}=0$ and $\phi_e = \phi$). Combining equations (8.22)–(8.27) then gives:

$$\frac{\mu}{\mu_0} K_0 = \left(\frac{K_b - K_{b\,min}}{K_{bm\,max} - K_{b\,min}} \right) \left[K_0 + \frac{\left(\dfrac{\rho_{ma} - \rho_b}{\rho_{ma} - \rho_f} \right)^{-1}}{1 - \left(\dfrac{\rho_{ma} - \rho_b}{\rho_{ma} - \rho_f} \right)\left(K_0 + \dfrac{\infty 4}{3}\mu_0 \right) - K_0^{-1}} \right] \qquad (8.28)$$

This one equation contains three unknowns, which are the density ρ_{ma} of the solid part of the rock and the incompressibility K_0 and shear modulus μ_0 of the solid part of the rock. The density ρ_f of the fluid is assumed to be 1 g/cc, while the bulk density ρ_b is known, because it is one of the products of the inversion.

Clearly, two more equations are needed to solve the system. These can be generated by first expressing the three unknown parameters as functions of $XCAL$, the volume fraction of calcite in the solid part of the rock, which is defined by:

$$XCAL = \frac{V_{CAL}}{V_{CAL} + V_{DOL}}, \qquad (8.29)$$

Since, by assumption, dolomite and calcite are the only minerals in the system, it is clear also that $XDOL = 1 - XCAL$.

The elastic parameters and bulk densities of the minerals calcite and dolomite have standard values that have been accurately determined (Table 8.6).

Two relationships between $XCAL$ and the elastic moduli can be established by using the Hashin-Shtrikman bounds. The upper and lower bounds of the moduli can be expressed as functions of $XCAL$, and if the averages of the bounds are taken the dependences of K_0 and μ_0 on $XCAL$ are fixed uniquely. For instance, in the case of K_0, the upper and lower bounds and the average are:

$$K_{0MAX} = K_{0DOL} + \frac{XCAL}{(K_{0CAL} - K_{0DOL})^{-1} + (1 - XCAL)\left(K_{0DOL} + \dfrac{4}{3}\mu_{0DOL} \right)^{-1}} \qquad (8.30a)$$

$$K_{0MIN} = K_{0CAL} + \frac{1 - XCAL}{(K_{0DOL} - K_{0CAL}) + XCAL\left(K_{0CAL} + \dfrac{4}{3}\mu_{0CAL} \right)^{-1}} \qquad (8.30b)$$

Table 8.6 Elastic parameters and bulk densities of the minerals calcite and dolomite

	K_0 (Gpa)	μ_0 (Gpa)	ρ_b (g/cc)
Calcite	70.8	30.3	2.71
Dolomite	80.2	48.7	2.88

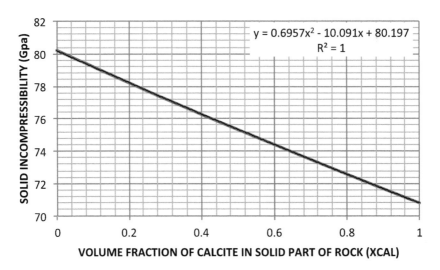

Figure 8.11 The relationship between XCAL and the incompressibility K_0 of the solid part of the rock, which is a mixture of calcite and dolomite.

$$K_0 = \frac{1}{2}(K_{0MIN} + K_{0MAX}) \tag{8.30c}$$

Figure 8.11 plots the relationship between K_0 and $XCAL$ given by equations (8.30a)–(8.30c).

The curve appearing in this figure can be fit almost exactly by a quadratic function, as shown in the upper right of the diagram. Furthermore, the upper and lower Hashin-Shtrikman bounds in this example are almost identical, so either of the two bounds could have been used just as easily to construct the quadratic fitting function.

A similar procedure can be applied to the shear modulus μ_0 of the solid part of the rock. The relevant equations are:

$$\mu_{0MAX} = \mu_{0DOL} + \cfrac{XCAL}{(\mu_{CAL} - \mu_{DOL})^{-1} + \cfrac{2(1-XCAL)(K_{DOL} + 2\mu_{DOL})}{5\mu_{DOL}\left(K_{DOL} + \frac{4}{3}\mu_{DOL}\right)}} \tag{8.31a}$$

$$\mu_{0MIN} = \mu_{0CAL} + \cfrac{1-XCAL}{(\mu_{DOL} - \mu_{CAL})^{-1} + \cfrac{2XCAL(K_{CAL} + 2\mu_{CAL})}{5\mu_{CAL}\left(K_{CAL} + \frac{4}{3}\mu_{CAL}\right)}} \tag{8.31b}$$

$$\mu_0 = \frac{1}{2}(\mu_{0MIN} + \mu_{0MAX}) \tag{8.31c}$$

Figure 8.12 shows the relationship between the shear modulus of the solid part of the rock and the volume fraction of calcite in the solid part of the rock, which again can be fit almost exactly by a quadratic function.

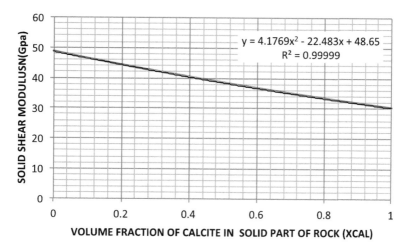

Figure 8.12 The almost perfect parabolic relationship between the shear modulus of the solid part of the rock and the volume fraction of calcite in the solid part of the rock.

It is a straightforward matter to write the matrix density ρ_{ma} as a function also of *XCAL*, as follows:

$$\rho_{MA} = \rho_{DOL} - XCAL(\rho_{DOL} - \rho_{CAL}) \tag{8.32}$$

So the three unknown parameters are all functions of *XCAL*, which is therefore effectively the only unknown in the problem.

To summarize, the equations for K_0, μ_0 and ρ_{ma} in terms of *XCAL* are:

$$K_0 = 0.6957 XCAL^2 - 10.091 XCAL + 80.197 \tag{8.33} \text{ (from Figure 8.11)}$$

$$\mu_0 = 4.1769 XCAL^2 - 22.483 XCAL^2 - 22.483 XCAL + 48.65$$
$$\tag{8.34} \text{ (from Figure 8.12)}$$

and

$$\rho_{ma} = 2.88 - 0.17 XCAL \tag{8.35} \text{ (from equation (8.32))}$$

If we replace K_0, μ_0 and ρ_{ma} in equation (8.28) with the expressions on the right of equations (8.33–8.35), what results is a cumbersome equation involving just the one unknown parameter, *XCAL*. This equation can be solved numerically using the Newton-Raphson method.

In practice, the chances of encountering a situation in which this methodology could be applied are slim. It is very unusual to come across a volume of rock that is completely devoid of shale or clay, and whose only components are fluid (water or/and oil), calcite and dolomite. Nonetheless, the case described here represents the "ideal" inversion, as the porosity and the mineralogical composition could in principle be estimated at any point in the seismic volume, without the need for well data.

REFERENCES

Connolly, P. (1999), Elastic impedance, *The Leading Edge*, 18, 4.

Davis, J. (1973), *Statistics and Data Analysis in Geology*, John Wiley & Sons, New York, Chichester, Brisbane, Toronto.

Lindsay, R. and Van Koughnet, R. (2001), Sequential backus averaging: Upscaling well logs to seismic wavelengths, *The Leading Edge*, 2001, 188–191.

Moss, B. (1997), The partitioning of petrophysical data, in Developments in Petrophysics, ed. by M. Lovell and P. Harvey, Geological Society Special Publication No. 122.

Spiegel, M. (1975), *Probability and Statistics, Schaum's Outline Seires*, McGraw-Hill Book Company.

Sveshnikov, A. (1968), *Problems in Probability Theory, Mathematical Statistics and Theory of Random Functions*, Dover Publications Inc., New York.

Veeken, P. and Da Silva, M. (2004), Seismic inversion methods and their constraints, *First Break*, 22, 6.

APPENDIX 8.1: MIXTURES OF NORMAL DISTRIBUTIONS

Quite frequently, when a histogram of GR values over a certain stratigraphic interval is plotted, a bimodal distribution of the GR values emerges. It is both tempting and reasonable to assume that the distribution is in fact a mixture of two independent distributions, each representing a separate "facies". Typically, the distribution with the larger mode corresponds to "shale" and the distribution with the smaller mode corresponds to "sand". If the two distributions can be completely separated, the facies problem is solved. However, we do not know *a priori* what the two distributions contributing to the mixture are. In what follows, we will assume that the two distributions are normally distributed. However, the methodology described below is quite general and could be applied to any other choice of distributions.

The probability density function $f(x)$ of a mixture of two distributions is:

$$f(x) = \alpha_1 f_1(x) + \alpha_2 f_2(x), \tag{8.1.1}$$

where $f_1(x)$ and $f_2(x)$ are the probability density functions of components 1 and 2, respectively, and α_1 and α_2 (with $\alpha_1 + \alpha_2 = 1$) are the relative proportions of the two components in the mixture.

If equation (8.1.1) is multiplied by x^n and integrated over all x, it is clear that nth moments of the distribution of the mixture are just the weighted sums of the nth moments of the distributions of the two components:

$$\int f(x) x^n \, dx = \alpha_1 \int f_1(x) x^n \, dx + \alpha_2 \int f_2(x) x^n \, dx. \tag{8.1.2}$$

Here, the first moment ($n = 1$) is the mean μ_1 of the distribution, while the second moment ($n = 2$) is often written as μ_2, and is equal to $\mu_1^2 + \sigma^2$, where σ^2 is the variance of the distribution. Similar interpretations apply to the first two moments of the two component distributions.

Furthermore, if both $f_1(x)$ and $f_2(x)$ are assumed to be normally distributed, the limits of integration in (8.1.2) will be $-\infty$ and ∞.

In the case of a mixture of two normal distributions whose statistical parameters are unknown, there are in total six unknown quantities (the two proportions, plus the mean and the standard deviation for each component). If the values of all six parameters were known, we would be able to calculate directly the probabilities that a point with a given value x of the random variable X belongs to facies 1 or 2.

If equation (8.1.2) is written out explicitly for the first five moments of the distribution, we have the system of six equations:

$$\mu_1 = \alpha_1\mu_{11} + \alpha_2\mu_{12} \tag{8.1.3a}$$

$$\mu_2 = \alpha_1\mu_{21} + \alpha_2\mu_{22} \tag{8.1.3b}$$

$$\mu_3 = \alpha_1\mu_{31} + \alpha_2\mu_{32} \tag{8.1.3c}$$

$$\mu_4 = \alpha_1\mu_{41} + \alpha_2\mu_{42} \tag{8.1.3d}$$

$$\mu_5 = \alpha_1\mu_{51} + \alpha_2\mu_{52} \tag{8.1.3e}$$

$$1 = \alpha_1 + \alpha_2 \tag{8.1.3f}$$

In these equations, μ_{jk} represents the jth moment of the kth component in the mixture. The moments of the whole distribution, on the left hand side of these equations, can be calculated directly from the measured values $(x_1, x_2, ..., x_N)$ of the discriminating variable, and are given by:

$$\mu_j = \frac{1}{N}\sum_{i=1}^{i=N} x_i^j, \tag{8.1.4}$$

where N is the total number of data points.

For a normal distribution, all the moments can be expressed as functions of the first two. The moments from the third onwards can be calculated by making use of the following moment-generating function (see, for example, Spiegel, 1975):

$$\mu_{jk} = \mu_{1k}\mu_{(j-1)k} + (j-1)(\mu_{2k} - \mu_{1k}^2)\mu_{(j-2)k} \tag{8.1.4}$$

For instance, the third moment of the kth population is given by:

$$\mu_{3k} = \mu_{1k}^3 + 3\sigma_k^2\mu_{1ka} \tag{8.1.5}$$

If the moment-generating function (equation 8.1.4) is used to eliminate the third, fourth and fifth moments in equations (8.1.3c)– (8.1.3e), what results is a system of six equations in six unknowns, which can be solved numerically using the Newton-Raphson method.

If data for two variables are available (taken, for example, from the GR and Bulk Density logs), and the distribution of the variables is assumed to be a mixture of two two-dimensional normal distributions, there are now 12 unknowns that would need to be determined before the component distributions can be completely separated.

The unknown parameters are, for each component in the mixture, the means and standard deviations of the two variables, plus the correlation coefficient between the two variables. With the addition of the two relative proportions, this gives 12 unknowns in total. The equivalent of the moment equation (8.1.2) in the case of two variables is:

$$\int\limits_{-\infty}^{\infty}\int\limits_{-\infty}^{\infty} f(x,y)x^n y^m \, dx\, dy = \int\limits_{-\infty}^{\infty}\int\limits_{-\infty}^{\infty} f_1(x,y)x^n y^m \, dx\, dy + \int\limits_{-\infty}^{\infty}\int\limits_{-\infty}^{\infty} f_2(x,y)x^n y^m \, dx\, dy \quad (8.1.6)$$

where the moments on the left side of this equation can again be calculated directly from the measured data. The higher order moments on the right side of equation 8.1.6 can always be calculated as functions of the first two moments, so in principle the problem can be formulated and solved in a manner analogous to the case with just one variable. However, in practice the case of two variables is far more cumbersome.

It should be mentioned that, in problems of this type, we are working with sample data rather than population data. Furthermore, the distributions underlying the measured data may be quite different from normal distributions, in which case the Newton-Raphson method typically yields no acceptable solution, because it never converges.

APPENDIX 8.2: SOME COMMENTS ON THE USE OF BAYES' THEOREM

Suppose that there are two possible events A_1 and A_2, whose probabilities $P(A_j)$ are complementary, so that $P(A_1) + P(A_2) = 1$. In terms of the applications of Bayes' theorem to facies identification described earlier in this chapter, event A_1 can be understood to mean that a certain point in a seismic volume corresponds to sand, and event A_2 that the point corresponds to shale. Consider also a third event, which we will call A, and is understood to mean that the value x of a certain geological variable lies in the interval:

$$x_0 \le x \le x_0 + \Delta x \qquad (8.2.1)$$

Then Bayes' theorem, as it is usually applied, allows us to estimate the probability $P(A_1|A)$ that the point corresponds to sand if it is known that the value x of the geological variable lies in the range given by (8.2.1). The relevant formula is:

$$P(A_1 \mid A) = \frac{P(A_1)P(A \mid A_1)}{P(A_1)P(A \mid A_1) + P(A_2)P(A \mid A_2)} \qquad (8.2.2)$$

where, in practical terms, $P(A_1)$ is the proportion of sand in the seismic volume under consideration – which is known – while $P(A_2) = 1 - P(A_1)$ is the proportion of shale, and $P(A|A_1)$ is the probability that the geological variable will lie in the range indicated in (8.2.1), given that the point is a sand. This last probability is also known, because the facies present in the seismic volume have already been identified, and the distributions of the geological variable in the individual facies have been estimated.

At this point, we would like to modify equation (8.2.2) so as to calculate the probability that a point in the seismic volume is sand, in the situation where the interval in equation (8.2.1) is infinitesimal.

This is easily done, and the required formula is:

$$P(A_1 \mid x) = \frac{P(A_1)f_1(x)}{P(A_1)f_1(x) + P(A_2)f_2(x)} \tag{8.2.3}$$

where, as before, $P(A_1)$ and $P(A_2)$ are the proportions of sand and shale in the volume under consideration, and f_1 and f_2 are the probability density functions of the geological variable in the two facies, evaluated at a point x, rather than over an interval as in equation (8.2.2). This formula has already been used above, in connection with inversion and AVO, and has also been extended, in equation (8.8), to the case of several variables.

Chapter 9

Modelling carbonates using Differential Effective Medium theory

INTRODUCTION

This chapter is somewhat different to the previous ones. In principle, the ideas presented here will not be directly applicable to inversion or AVO studies. The practical applications of this chapter belong more to the category of well log analysis. Furthermore, the applications are somewhat experimental. It has been noticed that the ideas work reasonably well in the very few cases in which they have been tested, but more work needs be done to confirm that the principles outlined here really work. The reader may ask: Why has this chapter been included at all? The answer is that the Differential Effective Medium (DEM) model, which is based on the simultaneous solution of two differential equations, enables us to predict the physical properties of carbonates, without the need for introducing empirical parameters. So, in a sense, it is the rock physicist's dream come true: a consistent theory of the elastic properties of carbonates. In this chapter we will attempt to (qualitatively) predict the permeability of carbonates, by applying DEM theory.

PRELIMINARY REMARKS

The porosity and mineralogical composition of carbonates (the proportions of calcite, dolomite and shale) can be calculated from the Density, Neutron and Gamma Ray logs. However, knowing the porosity alone is not enough to determine if a carbonate will produce hydrocarbons. It is quite common to encounter carbonates with porosities of 10% or more, which prove to be impermeable when tested. On the other hand, there are low-porosity carbonates which prove to be permeable and have a very good production of hydrocarbons. In order to be permeable, the pores have to be interconnected. In carbonates with a relatively high porosity, there may be isolated cavities or vugs, which account for the porosity, but the rock is impermeable because the vugs do not form a network. In a rock with low porosity, there may be cracks or fractures which connect many of the voids in the rock, resulting in a high permeability. Needless to say, the identification of a zone of permeable carbonates in a well is very important from an economic point of view.

In this chapter, we will outline some possible methods for identifying permeable zones in a well which make use of the porosity (as calculated from the radioactivity logs) and the sonic log (which measures the inverse of the velocity of the primary

DOI: 10.1201/9781003261773-10

waves). In particular, we will be dealing with rocks of a relatively low porosity (less than about 0.20). The basic idea is to model the primary velocity of the rocks as a function of the total porosity and compare the modelled results with the actual data, whenever this is possible. In order to model rocks perceived as permeable or impermeable, we will make use of the DEM model. A very thorough description of the DEM method is given by Berryman (1995) and Berryman et al. (2002). In this last reference, the authors solve the DEM equations analytically for penny cracks in saturated rocks. However, their solution is valid only for very low values of the aspect ratio (less than about 0.01). Mavko et al. (1998) also provide an excellent explanation of the method. These authors give a very succinct and clear explanation of the theory behind it, which we therefore quote verbatim:

> The DEM theory models two phase composites by incrementally adding inclusions of one phase (phase 2) to the matrix phase... The matrix begins as phase 1 (when concentration of phase 2 is zero) and is changed at each step as a new increment of phase 2 material is added. The process is continued until the desired proportion of constituents is reached. The DEM formulation does not treat each constituent symmetrically.

Anselmetti and Eberle (1999) observed that if the porosity of a set of carbonates (calculated from the radioactivity logs) is plotted against the P-wave velocity, the points have a large scatter. The Wyllie equation is generally a rough approximation to the observed data. The authors realized that the positions of the points in the diagram depend also on the pore type, and as a consequence developed the concept of the "velocity deviation log", using the Wyllie curve (expressed in terms of velocity) as a reference. Points with a large positive value of the deviation (i.e. for which the observed velocity is greater than the Wyllie velocity) correspond to moldic porosity. Points which fall in the vicinity of the Wyllie curve correspond to interparticle porosity. And points with a relatively small negative deviation correspond to microporosity. Fractured samples were excluded from the authors' study. In this chapter the concept of the velocity deviation log will be adopted, although in a somewhat different way. Note that in the Anselmetti and Eberli paper there is no reference to the DEM model.

It should be pointed out that the estimates generated by the method outlined here are qualitative rather than quantitative. In a crossplot of V_P against porosity, generated from the Neutron and Density logs, we should be able to identify points which represent vugs interconnected by fractures (an indication of permeable rocks) or isolated spherical vugs (an indication of impermeable rocks), provided that the solid part of the rock is either pure calcite or pure dolomite. However, while the end points are easily identifiable, there are intermediate points whose position in the diagram is ambiguous. One of the limitations of the DEM model, as will be seen later, consists in the fact that it is unable to model a granular medium; that is, if the original continuous phase is a fluid with a shear modulus equal to zero, and the final phase is constructed by adding mineral inclusions.

For instance, an oolite, or any carbonate with an intergranular and interconnected porosity, with its pores either dry or filled with a fluid, cannot be modelled using DEM theory. (In order to do this, we would need to resort to mathematical approximations, which rely on the introduction of constants that are very difficult to evaluate in practice, or alternatively, make use of hybrid models.) In very general terms, the DEM method starts from a body with zero porosity (like a block of cement), to which voids of different sizes and aspect ratios are added, so that at each step there is an infinitesimal change in the elastic properties of the body. Many carbonates can be characterized this way, and so are ideal candidates for the method.

Fractures will be modelled as ellipsoids with a small aspect ratio (less than 0.1). One of the predictions of this chapter is that, for small aspect ratios, the ellipsoids tend to be interconnected and hence generate a network resulting in a permeable rock. However, for higher aspect ratios (of the order of 0.1) we can reproduce the Willey equation quite closely. Empirical observations suggest that the points falling close to the Wyllie line represent intergranular or intercrystalline porosity (Anselmetti and Eberli, 1999). Rocks of this type tend to be permeable. On the other hand, rocks with substantial microporosity (which tend to be impermeable) also fall close to the Wyllie line, albeit a bit lower than those with intergranular or intercrystalline porosity. Hence, in the neighbourhood of the Wyllie line, we really do not know if a rock will be permeable or not. From DEM modelling, rocks with a relatively high aspect ratio should not be permeable. Hence, there is a fundamental ambiguity attached to the points falling close to the Wyllie line.

Despite these limitations, it is a worthwhile exercise to model carbonates using DEM theory, because in certain circumstances it gives accurate predictions of the hydrocarbon productivity of some intervals.

Here, we will model the following two cases:

1. The addition of spherical pores to a continuous carbonate matrix, which represent unconnected vugs and, by definition, model the pores in an impermeable rock.
2. The addition of ellipsoidal pores (with a low aspect ratio) to a rock with a continuous carbonate matrix and an initial vuggy porosity. The pores are assumed to be interconnected fractures, so that the final phase is a permeable rock.

THE CASE OF SPHERICAL PORES

The coupled system of ordinary differential equations that characterizes the DEM model is:

$$(1-\varphi)\frac{dK^*}{d\varphi} = \left(K_f - K^*\right)P \tag{9.1}$$

$$(1-\varphi)\frac{d\mu^*}{d\varphi} = -\mu^* Q \tag{9.2}$$

where K^* and μ^* are the bulk and shear moduli, respectively, K_f is the fluid bulk modulus and φ is the porosity. As shown here, equations (9.1) and (9.2) apply only when the host material is a solid matrix and the inclusions represent pores filled with a fluid with a bulk modulus equal to K_f. We will deal first with the case where the inclusions are spherical. For spherical inclusions, the functions P and Q are given by:

$$P = \frac{K* + \frac{4}{3}\mu*}{K_f + \frac{4}{3}\mu*}$$

(9.3)

$$Q = \frac{6 + \frac{(9K* + 8\mu*)}{(K* + 2\mu*)}}{\frac{(9K* + 8\mu*)}{(K* + 2\mu*)}}$$

(9.4)

After dividing equation (9.1) by equation (9.2) and introducing the expressions equations (9.3) and (9.4) for P and Q, we obtain the single differential equation for $K*$:

$$\frac{dK*}{d\mu*} = \frac{(K* - K_f)(9K* + 8\mu*)}{15\mu*\left(K_f + \frac{4}{3}\mu*\right)}$$

(9.5)

At this point, we need to solve equation (9.5) for $K* = f(\mu*)$, then replace $K*$ with $f(\mu*)$ in equation (9.1), and finally solve equation (9.1) to obtain $\phi = \Phi(\mu*)$. We can then easily evaluate V_P by means of the formula:

$$V_P = \sqrt{\frac{K* + \frac{4}{3}\mu*}{\rho_b}},$$

where the denominator in this expression is the bulk density, which is readily calculated.

The procedure therefore allows us to model V_P as a function of the porosity.

Let us first define three new variables $y = K*$, $x = \mu*$ and $K = K_f$. The details of the method used to solve equation (9.5) are given in Appendix 9.1, and the final result is:

$$\frac{y}{x} = \frac{K}{x} + \frac{x^{-\frac{2}{5}}(3K + 4x)^{-\frac{1}{5}}}{(G(x) + C)},$$

(9.6)

where C is an integration constant and $G(x)$ is shorthand for the integral:

$$G(x) = -\frac{9}{5}\int x^{-\frac{2}{5}}(3K + 4x)^{-\frac{6}{5}}dx$$

(9.7)

This integral can be represented in the form:

$$G(x) = \left(-\frac{9}{5}\right)4^{-1.2}\left[Ux^{0.6}{}_2F_1(0.6, 1.2, 1.6, Wx)\right]$$

(9.8)

where $_2F_1$ is the standard hypergeometric function, and the parameters U and W are defined below.

The parameters $a = 0.6$, $b = 1.2$ and $c = 1.6$ appearing in equation (9.8), as the arguments of the hypergeometric function, are always the same, independent of the value of K. However, the parameters U and W do depend on the value of K, the bulk modulus of the inclusions. If we define

$$T = \frac{3}{4}K,$$

it can be checked that

$$W = -\frac{1}{T}$$

(and so is always negative) and that

$$U = \frac{5}{3}T^{-1.2}$$

It should be noted that an alternative representation of equation (9.8), using different arguments for $_2F_1$, is the formula:

$$G(x) = \left(-\frac{9}{5}\right)4^{-1.2}\left[Ux^{0.6}(1-Wx)^{-1.2}{}_2F_1(1.2,1.0,1.6,\frac{Wx}{Wx-1})\right]$$

Equation (9.6) allows us to express the bulk modulus K^* as a function of the shear modulus μ^*, in the case of spherical inclusions filled with a fluid. However, the problem has not yet been solved completely, because we still need to express the porosity ϕ as a function of μ^* (or x, in the new set of variables). A suitable formula, found by integrating equation (9.1), is:

$$\ln\frac{1}{1-\phi} = \left(-\frac{3}{5}\right)\int \frac{K + \dfrac{x^{0.6}(3K+4x)^{-0.2}}{G(x)+C} + \dfrac{8}{9}x}{x\left[K + \dfrac{x^{0.6}(3K+4x)^{-0.2}}{G(x)+C} + \dfrac{4}{3}x\right]}dx + C'$$

where K is the bulk modulus of the fluid as before and C' is an integration constant. It is in almost all cases impossible to evaluate the expression on the right of this equation explicitly, and hence an empirical approximation will be used. However, as a by-product of this approach we will be able to estimate also the function $K^* = f(\mu^*)$, which will allow us to check the accuracy of the approximation against equation (9.6), which is the exact solution of the differential equation (9.5).

In the case of spherical inclusions, if we set $K = 0$ (which corresponds to the situation where the spherical pores are empty and $K^* = K_{dry}$), it is a straightforward matter to solve equations (9.1) and (9.2) simultaneously to give:

$$K^*(dry) = \frac{\frac{4}{3}\mu^*}{1 - \left(\frac{\mu^*}{C}\right)^{\frac{3}{5}}},$$

(9.9)

where C is an integration constant that is easily evaluated, and

$$1 - \varphi = \frac{\sqrt{\frac{\mu^*}{C}}}{\left[B\left[2 - \left(\frac{\mu^*}{C}\right)^{\frac{3}{5}}\right]^{\frac{1}{6}}\right]},$$

(9.10)

where B is a second integration constant, also easily evaluated.

To calculate $K^*_{(wet)}$ from $K^*_{(dry)}$, we proceed as follows. Note that in this case we cannot use the Gassmann equation for the calculations, because the rock is supposed to be impermeable.

As explained in Chapter 5, Hashin-Shtrikman theory allows us to calculate the maximum and the minimum possible values of any elastic modulus, provided we know the composition of the rock and the fluid filling the pore space. The true value of any elastic modulus lies between the upper and lower bounds. The true value M of any such elastic modulus can be therefore be expressed as:

$$M = wMHS^+ + (1-w)MHS^-,$$

(9.11)

where MHS^+ and MHS^- are the maximum and minimum possible values of the elastic modulus for the given composition, and the weight w is a number between 0 and 1. The maximum and minimum values can be calculated using formulas developed by Hashin and Shtrikman.

Marion's bounding average hypothesis, which was also introduced in Chapter 5, states that, for a given rock, the weight w is independent of the nature of the pore fluid. Hence, if we know the values of the modulus K^* for a dry rock, we can use equation (9.11) to calculate the weights for all possible porosities. Once the weight is known, we can use equation (9.11) to calculate K^* for a "wet" rock.

In the specific case of the bulk modulus K^* for a dry rock we have:

$$K_{dry} = wKHS^+_{dry} + (1-w)KHS^-_{dry}$$

(9.12)

where

$$KHS^+_{dry} = K_M + \frac{\varphi}{-K_M^{-1} + (1-\varphi)\left(K_M + \frac{4}{3}\mu_M\right)^{-1}},$$

and

$$KHS^-_{dry} = 0$$

where the subscript M on the bulk and shear moduli refers to the mineral constituting the solid part of the rock.

With K_{dry} calculated from equation (9.9), we can solve for w in equation (9.12) to give:

$$w = \frac{K_{dry}}{KHS_{dry}^+}$$

(9.13)

Finally, we calculate the bulk modulus of the wet rock by substituting equation (9.13) into equation (9.12), with MHS^+ and MHS^- on the right-hand side of the equation now replaced by the Hashin-Shtrikman upper and lower bounds:

$$KHS_{wet}^+ = K_M + \frac{\varphi}{(K_f - K_M)^{-1} + (1-\varphi)\left(K_M + \frac{4}{3}\mu_M\right)^{-1}}$$

$$KHS_{wet}^- = \frac{K_f K_M}{K_M \varphi + K_f (1-\varphi)}$$

The left-hand side of equation (9.12) is now understood to be $K_{(wet)}$, the bulk modulus of the fluid-saturated rock.

Once the value of bulk modulus K^* for the wet rock is known, we can use it to calculate the velocity of the primary (longitudinal) wave from the formula:

$$V_P = \sqrt{\frac{K^* + \frac{4}{3}\mu^*}{\rho_b}},$$

where ρ_b is the bulk density, and the value of the shear modulus μ^* is independent of the type of fluid inside the pores.

Figure 9.1 is a crossplot of the bulk modulus K^* against the shear modulus μ^*. The points shown in the diagram represent the exact solution given by equation (9.6), while the continuous line has been generated using the empirical method described above. Note that there is an almost perfect coincidence between the two curves. This very good match gives us the confidence that it is reasonable to apply the Marion hypothesis to calculate the saturated bulk modulus K^* from the dry bulk modulus K_{dry}.

Figure 9.2 plots values of V_P against the porosity ϕ, where the points represent data taken from a real well in which calcite is the most abundant mineralogical component, while the dotted line is the predicted DEM curve in the case of spherical inclusions, and the full line is the Wyllie empirical curve. The points correspond to a perforated interval of about 104 m, which had a production of more than 10,000 bopd. The porosity was calculated using the Neutron and Density logs. The large separation between the dotted line and the cloud of points is very conspicuous. It is clear that the interval would not be accurately modelled as a vuggy carbonate dominated by unconnected vugs. In fact, the high level of hydrocarbon production from the interval indicates that the pore space must be well connected.

Figure 9.3 shows a crossplot of velocity against porosity for an interval about 20 m thick in a carbonate sequence, whose mineral content is almost pure dolomite.

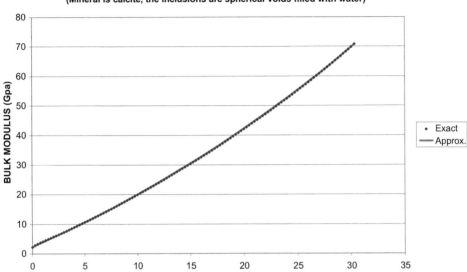

Figure 9.1 Comparison between the exact solution of the differential equations (the points) and the approximate solution (the solid line, calculated using the empirical method).

The relatively high porosities, which at many points are in the range of 0.15–0.25, made it an interesting prospect. However, the interval proved to be absolutely impermeable. It should be noted that the points having the maximum porosity tend to fall above the Wyllie line (solid), relatively close to the "spherical inclusions" dotted line. These points are interpreted to be indicative of unconnected vugs and hence impervious.

In summary, rocks with a vuggy or moldic porosity, where the pores are presumably not connected, can be modelled using spherical inclusions. The points will generally fall above the Wyllie line in a crossplot of velocity against porosity and should be easily identifiable. This expectation agrees with the empirical observations of Anselmetti and Eberle (1999). It should be pointed out that, if the pores are not spherical but instead have a relatively high aspect ratio (*i.e.* greater than about 0.1), the DEM curve will move "downwards" (as will be seen in the next section). However, as suggested in Appendix 9.3, if the pores have an aspect ratio of about 0.1, the rock will still be fairly impermeable, as are many of the points shown in Figure 9.3, which are situated below the purple line but far above the Wyllie line.

As for the points falling close to the Wyllie line (either above or below it), it is not possible, within the confines of the DEM model, to predict whether they will produce or not. This problem will be discussed later.

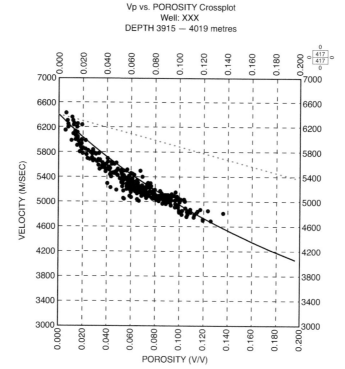

Vp vs. POROSITY Crossplot
Well: XXX
DEPTH 3915 — 4019 metres

Functions:
DEM (SPHERES IMMERSED IN A SOLID MATRIX)
EMPIRICAL WILEY FORMULA TO PREDICT POROSITY FROM Vp.

Figure 9.2 P-velocity vs. porosity for a clean carbonate with a high proportion of calcite. The curve which passes very close to the data points is the Wyllie curve. The dotted line above corresponds to a vuggy carbonate with non-connected vugs (an impervious rock). This interval has proved to be very permeable and the actual data points lie a long way from the curve for vuggy inclusions, as they should.

Figure 9.4 shows a crossplot of the bulk modulus of a wet rock against the porosity. The curve generated by the DEM method assuming spherical inclusions filled with a fluid has been plotted jointly with the Hashin-Shtrikman upper bound for this bulk modulus $K_{(wet)}$. Note that, for relatively low porosities, these two quantities are practically the same. The mineral matrix is pure calcite in both cases.

Figure 9.5 is a plot of the velocity against the porosity in the case of fluid-filled spherical inclusions, as well as the maximum possible velocity (calculated using the Hashin-Shtrikman upper bounds for the two elastic constants) plus the Wyllie line. As expected, the predicted velocity is very close to the maximum possible velocity, at least for relatively low porosities.

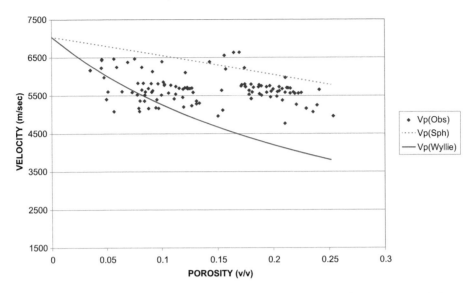

Figure 9.3 P-velocity vs. porosity for an impermeable interval of almost pure dolomite. The solid line is the Wyllie line. Note that the majority of the points lie quite above the Wyllie line, relatively close to the dotted line, which corresponds to spherical inclusions.

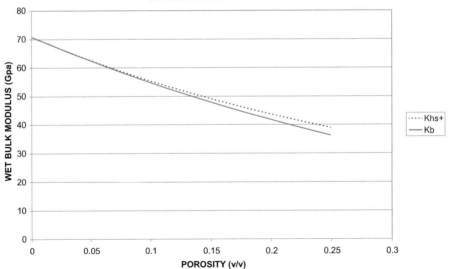

Figure 9.4 Bulk modulus vs. porosity for a calcite carbonate containing spherical pores or vugs. The predicted curve for vuggy inclusions coincides with the Hashin-Shtrikman upper bound for porosities of up to about 10%.

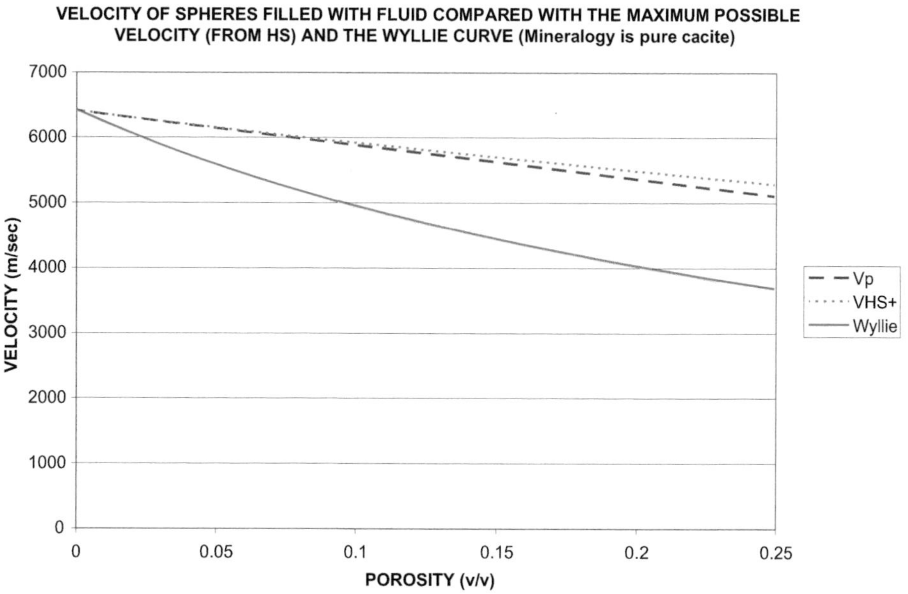

VELOCITY OF SPHERES FILLED WITH FLUID COMPARED WITH THE MAXIMUM POSSIBLE
VELOCITY (FROM HS) AND THE WYLLIE CURVE (Mineralogy is pure cacite)

**VELOCITY OF SPHERES FILLED WITH FLUID COMPARED WITH THE MAXIMUM POSSIBLE
VELOCITY (FROM HS) AND THE WYLLIE CURVE (Mineralogy is pure cacite)**

Figure 9.5 Velocity vs. porosity for a pure limestone containing spherical pores or vugs (the dashed line) and the velocity taken from the H-S upper bounds for the moduli (the dotted line). For the purposes of comparison, the Wyllie line has been included.

THE CASE OF PENNY CRACKS (REPRESENTING FRACTURES)

Penny cracks are very flat ellipsoids, which have two equal and very large axes, relative to the remaining axis. The aspect ratio is the quotient between the short and the long axis. Of course, there is a continuum between a sphere and a penny crack. As an arbitrary convention, we can consider a "penny crack" to be any ellipsoid that has an aspect ratio less than 0.1. The formulas presented in this section are valid for values of the aspect ratio less than about 0.2.

The pores in this case are assumed to be empty. To simplify the notation, we will let K denote the dry bulk modulus, μ the shear modulus, α the aspect ratio of the ellipsoids making up the penny cracks and y the porosity of the rock.

In this section we will calculate K and μ as functions of the porosity y for dry rocks. We will have to calculate the saturated bulk modulus, following the empirical method used in the previous section.

The procedure outlined here starts from a rock with "matrix" porosity, which can be constructed using the DEM model with spherical inclusions. The penny cracks represent the fracture porosity that is present in addition to the matrix porosity. Although the pores in the original phase are not interconnected, they should be interconnected after the addition of the penny cracks.

After dividing the differential equations (9.1) by (9.2), we get:

$$\frac{dK}{d\mu} = \left(\frac{K}{\mu} \right) \frac{P}{Q}, \tag{9.14}$$

where

$$P = \frac{K(3K+4\mu)}{\alpha\pi\mu(3K+\mu)}$$

$$Q = \frac{(9K+6\mu)(3K+\mu)\alpha\pi + 4(3K+4\mu)(9K+4\mu)}{5\alpha\pi(9K+6\mu)(3K+\mu)}$$

On introducing the variable $v = K/\mu$ and making a few algebraic manipulations, we find that:

$$\frac{d\mu}{\mu} = \frac{27v^2(4+\alpha\pi) + 3v(64+9\alpha\pi) + (64+6\alpha\pi)}{135v\left[v^3 + \left(\frac{162-27\alpha\pi}{135}\right)v^2 - \left(\frac{72+27\alpha\pi}{135}\right)v - \left(\frac{64+6\alpha\pi}{135}\right)\right]} dv \qquad (9.15)$$

Details of the integration of equation (9.15) are given in Appendix 9.2. Once the equation has been integrated and the porosity has been expressed as a function of K/μ, it is possible to express μ in the form:

$$\mu = Cv^{A_1}(v-\rho_1)^{A_2}(v-\rho_2)^{A_3}(v-\rho_3)^{A_4}, \qquad (9.16)$$

where C is an integration constant, the A's and the ρ's are known functions of the aspect ratio, and:

$$-\ln(1-y) = B_0 \ln v + B_1 \ln(v-\rho_1) + B_2 \ln(v-\rho_2) + B_3 \ln(v-\rho_3) + C', \qquad (9.17)$$

In equation (9.17), the B's are also known as functions of the aspect ratio, while C' is a new integration constant.

Equations (9.16) and (9.17) are valid for all aspect ratios less than 0.20. For values of the aspect ratio greater than this, the predicted value of μ exceeds the Hashin-Shtrikman upper limit for the shear modulus.

Before considering the case of penny cracks – which correspond to aspect ratios less than about 0.1 – it is instructive to consider the situation in which the aspect ratio is greater than about 0.1. Figure 9.6 shows a crossplot of velocity against porosity, with the curve predicted by the DEM model with spherical inclusions – described by the formulas in the previous section – (dots), the curve corresponding to the addition of penny cracks with an aspect ratio of 0.1 – described by equations (9.16) and (9.17) – (dashes), and the Wyllie line (solid). Given that spheres have an effective aspect ratio of 1, it is evident that the effect of a decreasing aspect ratio is to move the velocity response curve downwards. For an aspect ratio equal to 0.1, the predicted curve almost coincides with the Wyllie line. What would the implications be for the permeability of the rocks at points falling close to the Wyllie line, according to the "pure" DEM model?

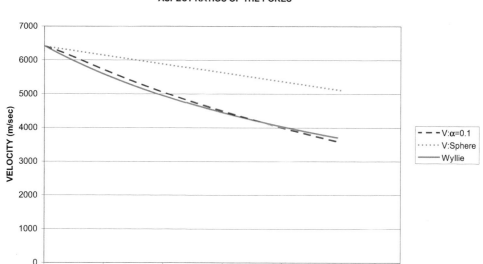

CHANGES IN THE VELOCITY vs. POROSITY CURVES AS A FUNCTION OF THE ASPECT RATIOS OF THE PORES

Figure 9.6 P-Velocity vs porosity for a pure limestone containing pores with two aspect ratios: 1 and 0.1. The Wyllie curve solid and almost coincides with the curve for the case with an aspect ratio 0.1.

As shown in Appendix 9.3, a rock whose pores have an aspect ratio of about 0.1 is quite likely to be impermeable, as the chance that the pores form a connected network is low. DEM theory suggests that, whenever the observed points fall close to the Wyllie line, it is quite likely that the rock would not produce hydrocarbons. Empirical observations suggest that points falling close to the Wyllie line correspond to interparticle porosity (Anselmetti and Eberle, 1999). Depending on the size of the pores and the degree of cementation of such a rock, it may or may not produce hydrocarbons. The problem is that according to "pure" DEM theory (*cf.* Appendix 9.1), it is not possible to simulate a granular medium in which the continuous phase consists of pores filled with a fluid, and hence we cannot check this empirical observation. In fact, in order to simulate a granular medium one has to resort to other approaches, such as the Hertz-Mindlin model. One could imagine starting with a granular rock with a given porosity, whose bulk and shear moduli are determined by Hertz-Mindlin theory, then adding infinitesimal amounts of cement to the granular medium, in line with the DEM method. Because the Hertz-Mindlin model for a pure granular medium generates a velocity against porosity curve that lies well below the Wyllie curve, one would expect that the addition of cement would move the curve "up" towards the Wyllie line. A hybrid model of this type is beyond the scope of this book. For the moment, the ambiguous nature of the permeability of points falling close to the Wyllie line remains unresolved, and different models could in principle give predictions that are diametrically opposed.

Consider now the case of penny cracks, where the aspect ratio is less than 0.1. It is generally accepted that the actual fracture porosity is a relatively small quantity, rarely exceeding 2%. However, the rock always has an additional porosity, either intergranular or vuggy. We will assume that the other porosity is vuggy and will model it using the DEM model for water-filled spheres. Altogether, the whole system of pores is presumed to form a connected network, in which the penny cracks intersect each other and the vugs. The resulting rock should be very permeable.

Figure 9.7 shows a crossplot of K_b against porosity for a rock with a vuggy porosity of 0.04, plus a penny crack porosity with an aspect ratio of 0.001. The mineralogy is pure calcite. Note that when the fracture or penny crack porosity exceeds about 1%, the curve asymptotically approaches the Hashin-Shtrikman lower bound.

Figure 9.8 is conceptually similar to the previous one, but the vertical axis measures the velocity rather than the bulk modulus. The lower bound for the velocity corresponds to a bulk modulus equal to the Hashin-Shtrikman lower bound, while the shear modulus, which has been set to 0, lies on the Hashin-Shtrikman lower bound as well.

The important feature of this plot is that for a significant fracture porosity, in the order of 1%, and a very low aspect ratio of the penny cracks, the velocity closely approaches the minimum possible velocity, as determined by the Hashin-Shtrikman bounds. Provided that the mineralogy is fairly constant over the interval of interest, it should be possible to identify from crossplots those points representing penny cracks

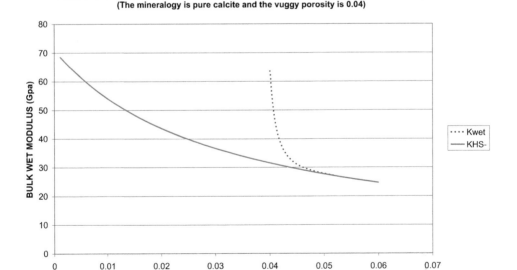

Figure 9.7 Asymptotic approach to the Hashin-Shtrikman lower bound for a rock containing penny cracks with an aspect ratio of 0.001. The solid line represents the H-S lower bound for the wet bulk modulus of limestone.

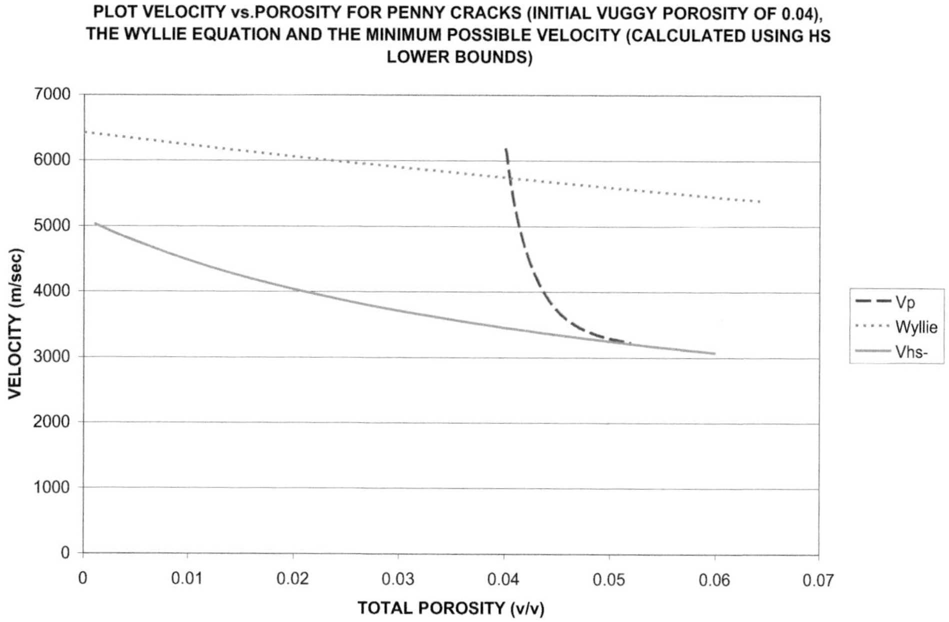

PLOT VELOCITY vs.POROSITY FOR PENNY CRACKS (INITIAL VUGGY POROSITY OF 0.04), THE WYLLIE EQUATION AND THE MINIMUM POSSIBLE VELOCITY (CALCULATED USING HS LOWER BOUNDS)

Figure 9.8 P-velocity vs. porosity for a limestone with 0.04 vuggy porosity and additional porosity due to penny cracks with an aspect ratio of 0.001. Note that V_P asymptotically approaches the minimum possible velocity curve.

with a significant fracture porosity. Note, however, that the defining feature of these points is that they fall below the Wyllie line for fracture porosities of interest, which is to say for the porosities in the order of 0.005.

It should be pointed out that the Hertz-Mindlin model for uncemented granular media produces a curve quite close to the velocity curve derived from the Hashin-Shtrikman lower bounds. However, a granular rock without cement (in which the grains are held together by compaction only) will generally be very permeable, unless the grains and pores are very small. So, although there is some uncertainty, any points in a velocity versus porosity diagram for a rock with a more or less constant composition, which falls close to the minimum possible velocity derived from the Hashin-Shtrikman lower bounds, should be permeable, either because of the presence of penny cracks with a small aspect ratio or because there is an interconnected network of pores, as is typical of an uncemented, granular medium.

Some comments are in order about the definition of the "minimum possible velocity". The velocity depends on both the bulk and the shear moduli. We have defined "the minimum possible velocity" here to be the velocity that results when two elastic moduli are equal to their Hashin-Shtrikman lower bounds. The Hashin-Shtrikman lower bound for K_b is simply:

$$KHSlow = \frac{K_m K_f}{K_m \varphi + K_f (1 - \varphi)},$$

where K_m and K_f are the bulk moduli of the mineral and the fluid components, respectively.

For the shear modulus, we have:

$$\mu HSlow = \mu_f + \frac{5\mu_f(1-\varphi)(\mu_m - \mu_f)(K_f + \frac{4}{3}\mu_f)}{5\mu_f(K_f + \frac{4}{3}\mu_f) + 2\varphi(K_f + 2\mu_f)(\mu_m - \mu_f)},$$

where, as before, the subscript "f" refers to the fluid while "m" refers to the mineral fraction. Note that the shear modulus of a fluid is zero, so in this case $\mu HSlow$ is generally zero. However, for the very specific case when the porosity is zero, the above expression indicates that the Hashin-Shtrikman lower bound for the pure mineral fraction is equal to the shear modulus of the mineral itself. So there is a discontinuity in the value of the minimum possible velocity when the porosity is equal to zero.

Figure 9.9 shows a crossplot of velocity against neutron porosity, for a 40 m interval of a carbonate composed mainly of calcite, free of shale. This interval was tested

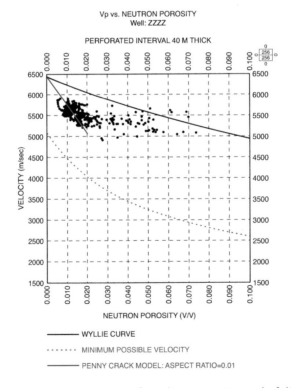

Figure 9.9 P-Velocity vs. neutron porosity for a limestone interval of 40 m thick, which is hydrocarbon-producing. The thin, solid line corresponds to a limestone with an aspect ratio of 0.01. The upper line is the Wyllie curve.

and produced about 5100 bopd, with a 3/8" choke. Note that most of the points in Figure 9.9 fall well below the Wyllie curve, while the porosity is quite low (the average neutron porosity is 0.022). It is apparent that despite the low porosity, the permeability is quite good, so the assumption that this interval is somewhat fractured appears to be reasonable. At least the points occupy the region where they would be expected to be if there were fractures, according to DEM theory.

In this figure, the line corresponding to penny cracks with an aspect ratio of 0.01 is shown as a thin, solid line. This line has been constructed on the assumption that all the porosities correspond to the penny cracks, and so there is no additional inter-granular or vuggy porosity. Note that calculations have been performed to a porosity of 2% only, because it is expected that the fracture porosity will never be greater than this. In this particular example, the total porosity measured by the horizontal axis of Figure 9.9 is equal to the fracture porosity.

DISCUSSION

According to the theory outlined above, it should be possible to detect permeable zones in a column of carbonates by using the following procedure, provided that the mineralogy of the column is more or less constant.

Given a crossplot of velocity against porosity, we can identify the following zones:

1. Points that are below the velocity curve corresponding to spherical inclusions filled with a fluid (for porosities up to 10%, this curve coincides almost with the maximum possible velocity, as calculated from the Hashin-Shtrikman upper bounds), but are above the Wyllie curve, should be impermeable according to DEM theory.
2. Points that are located in the neighbourhood of the Wyllie curve cannot definitively be classified as permeable or impermeable, according to "pure" DEM theory.
3. Points which fall below the Wyllie curve, but are relatively close to the curve of minimum possible velocity, should be permeable according to the DEM theory for penny cracks. Note that the use of the Hertz-Mindlin model for granular me-dia would place the points in similar positions, but most likely the corresponding rocks would also be permeable.

Figure 9.10 is a conceptual diagram in which all these zones have been marked. The calculations on which this diagram is based assume that the mineral has the properties of calcite.

Figure 9.11 is analogous to the previous one, but here curves generated on the as-sumption that the mineral has the properties of dolomite are plotted together with the curves for calcite. The purpose of this plot is to convey the influence the lithology has on the boundaries of zones with different permeabilities.

It is apparent from Figure 9.11 that changes in the mineralogy add further ambi-guity to the question of whether the rock is expected to be permeable at any point on the diagram, particularly in the zone of low permeability. So, in the case where the carbonate is a mixture of calcite and dolomite whose actual proportions are known, the following procedure for identifying the permeable and impermeable zones is

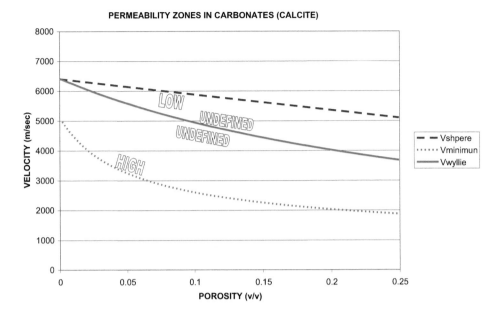

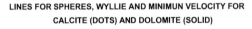

Figure 9.10 A conceptual diagram of P-velocity vs. porosity, illustrating the fact that zones with spherical vugs tend to be impermeable (aspect ratio = 1), while points close to the Wyllie line (aspect ratio close to 0.1) are of undefined permeability. The lowest line is the minimum possible velocity line.

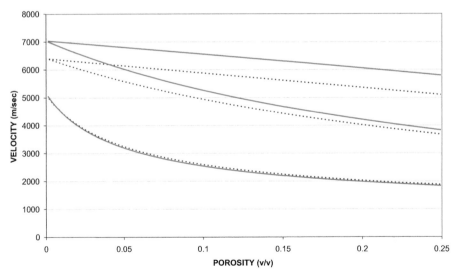

Figure 9.11 Complications involved in the definition of the permeable zones if the mineralogical composition of the rock is not known.

suggested. Since we are working with bulk moduli rather than with velocities, a shear wave velocity log should be available. The procedure also takes into account the properties of the fluid filling the pore space.

We can make the reasonable assumption that, in a rock whose observed bulk modulus for a particular porosity and composition is equal to the theoretical bulk modulus predicted by the DEM model for spherical inclusions filled with fluid, the probability that the rock will be permeable is zero. Similarly, when the observed bulk modulus for a given porosity and composition is equal to the value given by the 0.1 aspect ratio curve, the probability that the rock will be permeable is 0.5 (as the uncertainty in the permeability is maximum at that point). Finally if the observed bulk modulus is equal to the minimum bulk modulus (the Hashin-Shtrikman lower bound), the probability that the rock is permeable should be 1. For a given value of the observed bulk modulus, one can interpolate linearly to calculate the probability that the rock will be permeable at that point. Points with a significant proportion of shale (more than 5%, say) should be excluded from the analysis. For each point in the column we can generate a curve representing the probability that the rock will be permeable at that point, which can be plotted as a function of depth. This curve could act as a first filter to eliminate points, which with a high certainty will not produce hydrocarbons, and to pick out those intervals that are more likely to produce. The following example illustrates the procedure.

Figure 9.12 plots the bulk incompressibility against the porosity for a carbonate column about 100 m thick. The mineralogy of the column is a mixture of calcite and dolomite. As a visual aid, the curve corresponding to the maximum possible value of K_b (*i.e.* when the mineral fraction is 100% dolomite, and the pore fluid is water) is shown as dots, while the minimum possible value of K_b (corresponding to 100% calcite, and oil and gas and water in the pore space) is shown as a solid line. In fact, since each point has a different mineralogy and a different fluid content, the maximum and minimum possible curves given by the Hashin-Shtrikman bounds will be different. Hence, it is rather difficult to use this plot to determine if the rock is permeable at any particular point.

At this stage, we calculate the probability, at each point in the column, that the rock in the well will be permeable. At each point we know the porosity, the mineralogical composition and the fluid content. With these parameters we can calculate the Hashin-Shtrikman upper and lower bounds for K_b. Also, using these parameters, together with equations (9.16) and (9.17) (*cf.* Appendix 9.2) we can calculate K_b for an aspect ratio of 0.1. By assumption, the probability of production is equal to 0 if the observed K_b is equal to K_{bmax}, 0.5 if the observed K_b is equal to the value of K_b calculated for an aspect ratio of 0.1 and 1 if the observed K_b is equal to the Hashin-Shtrikman lower bound. From these three data points, we can estimate the probability of production for the observed value of K_b and ultimately generate a curve of the probability of hydrocarbon. Such a curve is included in Figure 9.13 (in the rightmost track) and corresponds to the data from the crossplot.

For this particular case, the curve yields poor results, as the production probability of most of the points is below 0.50. There are a few small intervals where the probability is greater than 0.5 (two of these intervals are within the perforated interval, which produced oil at a moderate rate), but in reality it would be extremely difficult to make a decision to perforate an interval based on this probability curve. In fact, the

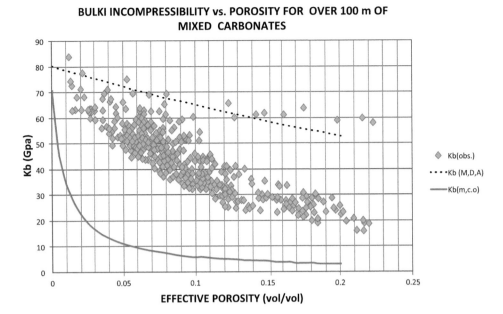

Figure 9.12 K_b vs. porosity diagram for to a column of mixed carbonates about 100 m thick. The curves are visual aids representing the maximum possible value of K_b (dotted line, where M stands for Maximum, D for dolomite and A for water) and the minimum possible value of K_b (solid line, where m stands for minimum, c for calcite and o for hydrocarbons plus water). Points with $V_{sh} > 5\%$ have been excluded. Points above the dotted line are not feasible and probably include an extraneous mineral.

resistivity curve was used to define the interval to be perforated (*cf.* Figure 9.13). This example does not help to confirm whether the method works or not. It is included here because the data are very complete, as all the well logs (including the shear wave log) are available and there is also a PVT. The reservoir was below the bubble point when the well was drilled.

It should be borne in mind that the probability curve in Figure 9.13 does not represent a genuine probability. Instead, it is a diagnostic aid used to determine if the bulk modulus at a point in the log falls above or below the 0.1 aspect ratio curve, in a plot of incompressibility against porosity.

FINAL REMARKS

In this chapter, we have described, with the aid of examples, how permeable zones in carbonates may be identified using the DEM model. DEM theory allows for the identification of the most impermeable points (which are modelled as spherical cavities filled with a fluid in an otherwise continuous solid phase, and interpreted as vuggy or moldic pores) and the most permeable points (which are modelled as penny cracks or ellipsoids with a low aspect ratio, and interpreted as fractures).

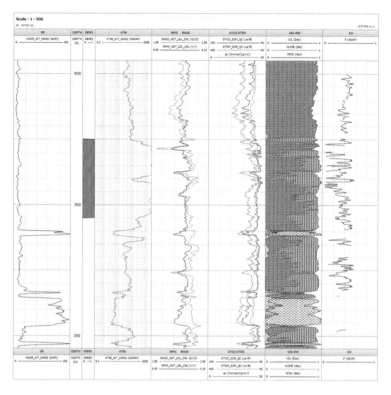

Figure 9.13 The rightmost track shows the probability curve calculated for the points in the 100 m interval. Points where this probability is greater than 0.50 have been shaded. The grey rectangle in the third track from the left corresponds to the perforated interval, which produced oil at a moderate rate. Note that the first digit of the depth measurements has been erased.

The theory gives ambiguous predictions for points which fall close to the Wyllie curve in a velocity versus porosity diagram. A model assuming elliptical pores with an aspect ratio of about 0.1 closely matches the Wyllie curve, and if reliable would imply that the points are more or less impermeable, as the connectivity of the pores is inversely proportional to the aspect ratio (*cf.* Appendix 9.3). However, empirical observations made by Anselmetti and Eberli suggest that points falling along the Wyllie line have interparticle porosity (in the most general sense) and these points may or may not be permeable, depending on the size of the pores and the amount of cement filling the pore space.

It is not possible within the scope of DEM theory to model a rock consisting of solid spheres surrounded by an otherwise continuous fluid phase. In other words, the theory cannot describe a medium with interparticle porosity and cannot be used to check Anselmetti and Eberli's observations. This failing can be demonstrated through numerical examples but will be given a rigorous justification in Appendix 9.1. The analytical solution of the DEM equations, presented in this chapter, provides a further insight into the strengths and limitations of the method. In order to model granular media, one has to rely on other approaches, such as the Hertz-Mindlin model, which could in principle be combined with DEM theory. A hybrid theory of this type could possibly reproduce the Wyllie curve, but the ambiguity in classifying the permeability of the points would remain.

The generalizations proposed here about the zones of high and low permeabilities need to be checked more extensively. The database that has been used in the examples presented so far is limited, and comparisons of many more production tests with the corresponding well log data would be desirable. As the permeability of the points falling along the Wyllie line cannot be predicted within the scope of DEM theory, it would be interesting to examine the productions tests corresponding to points of this type, to determine what their permeability is in practice.

REFERENCES

Anselmetti, F. and Eberle, G., (1999), The velocity deviation log: A tool to predict pore type and permeability trends in carbonate drill holes from sonic and porosity or density logs, *AAPG Bulletin*, 83, 3, 450–466.

Berryman, J.G, (1995), Mixture Theories for rock properties, in *Handbook of Physical Constants*, T.J. Ahrens, ed., American Geophysical Union, Washington D.C., 205–228.

Berryman, J.G., Pride, S.R. and Wang, H., (2002), A differential scheme for elastic properties of rocks with dry or saturated cracks, *Geophysical Journal International*, 151, 597–617.

Mavko, G., Mukerji, T. and Dvorkin, J., (1998), *The Rock Physics Handbook*, Cambridge University Press.

APPENDIX 9.1: EXACT SOLUTION OF THE DEM DIFFERENTIAL EQUATION IN THE CASE OF SPHERICAL INCLUSIONS FILLED WITH FLUID

The impossibility of simulating a granular medium using "pure" DEM theory

In the case of spherical inclusions – without any assumptions at this stage about the nature of the inclusions, which could be fluid-filled pores inside a solid or (in the granular case) solid spheres inside a fluid – the DEM differential equation can be expressed in the form

$$\frac{dy}{dx} = \frac{1}{5} \frac{(K-y)\left(6My + 12Mx + 9xy + 8x^2\right)}{x(M-x)(3K+4x)}, \tag{9.1.1}$$

where K and M are the bulk and shear moduli of the inclusions, respectively, and y and x are the bulk and shear moduli of the two-phase composite.

Equation (9.1.1) can be rewritten schematically as

$$\frac{dy}{dx} = ay^2 + by + c, \tag{9.1.2}$$

where

$$a = -3(2M + 3x)/[5(M - x)x(3K + 4x)]$$

$$b = (6KM + 9Kx - 12Mx - 8x^2)/\left[5(M - x)x(3K + 4x)\right]$$

and

$$c = 4K(3M + 2x)/[5(M - x)(3K + 4x)]$$

Equation (9.1.2) is a Ricatti differential equation which, in certain instances, can be solved analytically.

It is clear from (9.1.1) that one solution of the equation is just $y = K$. If we therefore write:

$$y = K + 1/u, \tag{9.1.3}$$

where u is a function to be determined, then substituting equation (9.1.3) into equation (9.1.2) gives:

$$-\frac{\dfrac{du}{dx}}{u^2} = \frac{a}{u^2} + \frac{(2Ka + b)}{u},$$

or equivalently

$$\frac{du}{dx} + pu = q, \tag{9.1.4}$$

where

$$p = 2Ka + b$$

and

$$q = -a$$

The general solution of the linear equation (9.1.4) is:

$$u = \frac{1}{h}\int hq\,dx + C,$$

where

$$h = \exp\left(\int p\,dx\right)$$

and C is an integration constant.

Now, p has the partial fractions decomposition

$$p = -\frac{2}{5x} - \frac{1}{M-x} - \frac{\frac{4}{3K+4x}}{5}$$

and so:

$$\int p\,dx = -\frac{2\ln(x)}{5} - \ln(M-x) - \frac{\ln(3K+4x)}{5}$$

Hence:

$$h = x^{-\frac{2}{5}}(M-x)(3K+4x)^{-\frac{1}{5}}$$

and therefore:

$$hq = \frac{3}{5}(2M+3x)x^{-\frac{7}{5}}(3K+4x)^{-\frac{6}{5}}$$

If we now define

$$G = \int hq\,dx = \frac{3}{5}\int(2M+3x)x^{-\frac{7}{5}}(3K+4x)^{-\frac{6}{5}}\,dx$$

then the general solution of equation (9.1.4) is:

$$u = \frac{x^{\frac{2}{5}}(3K+4x)^{\frac{1}{5}}(G+C)}{(M-x)},$$

where C is an integration constant.

The general solution of the original Ricatti equation (9.1.1) is therefore:

$$y = K + \frac{(M-x)x^{-\frac{2}{5}}(3K+4x)^{-\frac{1}{5}}}{G+C} \qquad (9.1.5)$$

As stated above, equation (9.1.5) is very general and in principle should be reliable in two particular cases:

a. If the original continuous medium is solid, and spherical fluid-filled inclusions are added, to model the presence of isolated vugs. In this case, the shear modulus M of the inclusions is zero and K is the bulk modulus K_f of the fluid.

b. If the original continuous medium is a liquid, to which spherical grains of solid mineral are added. The resulting medium will consist of a continuous network of pores and is granular. In this case, $M>0$ is the shear modulus of the mineral fraction and K is its bulk modulus.

In order to analyse the behaviour of the solutions of the Ricatti equation, we will first consider the asymptotic form of the integral

$$G(x) = \frac{3}{5}\int \frac{2M+3x}{x^{\frac{7}{5}}(3K+4x)^{\frac{6}{5}}}dx$$

for small values of x. Since

$$\frac{3}{5}\frac{2M+3x}{x^{7/5}(3K+4x)^{6/5}} \cong \frac{2}{15}3^{4/5}MK^{-6/5}x^{-7/5} - \frac{1}{75}3^{4/5}(16M-15K)K^{-11/5}x^{-2/5}$$

when $x \cong 0$, it follows that

$$G(x) \cong -3^{-1/5}MK^{-6/5}x^{-2/5} - \frac{1}{45}3^{4/5}(16M-15K)K^{-11/5}x^{3/5}$$

(plus an integration constant, which can be absorbed into C in equation (9.1.5)). This means that, if $M > 0$, the behaviour of the solution y in equation (9.1.5) for small values of x is:

$$y(x) \cong -3^{1/5}M^{-1}CK^{11/5}x^{2/5}$$

Clearly, there is no value of the constant C that will satisfy the conditions for an initial fluid phase, which are that $x=0$ and $y=K_f \neq 0$. In fact, whenever $M>0$, it is clear that the bulk modulus y must be 0 if the shear modulus x is 0.

On the other hand, if $M = 0$ the behaviour of the solution y in equation (9.1.5) for small values of x is:

$$y(x) \cong K + 3^{-1/5}C^{-1}K^{-1/5}x^{3/5}$$

and again the value of C does not affect the value of y in the limit $x \to 0$. But this is not a problem in the case where the inclusions are fluid-filled pores, because the limit $x \to 0$ then corresponds to the situation in which the entire solid matrix has been replaced by fluid, and $y(0) = K$ is the correct physical limit.

In summary, no physically acceptable solution $y(x)$ of the DEM equations exists if $M>0$, which entails that the medium must be initially solid (*i.e.*, the shear modulus of the inclusions must be zero).

APPENDIX 9.2: INTEGRATION OF THE DEM EQUATIONS IN THE CASE OF PENNY CRACKS, WHEN THE PORE SPACE IS EMPTY (*I.E.* DRY)

The notation used here is the same as in the body of the chapter. That is, K is the dry bulk modulus, μ is the shear modulus, α is the aspect ratio of the ellipsoids making up the penny cracks (which is assumed to be <<1) and y is the porosity of the rock.

In what follows, K and μ will be calculated as functions of the porosity y on the assumption that the rocks remain dry (*i.e.* the pore space is empty). It is assumed that the penny cracks form an interconnected network and hence, once K has been calculated, the bulk modulus for the corresponding fluid-saturated rock can be determined using either the Gassmann equation (which is valid only if the pores form an interconnected network), or the procedure outlined in the main part of this chapter.

The initial phase of the rock is assumed to have "matrix" porosity, which corresponds to a DEM model with spherical inclusions. The fracture porosity that is present in addition to the "matrix" porosity is then represented by penny cracks. Although the pore space is initially unconnected, once the penny cracks are added it should become connected.

As explained above, it is possible to rewrite the differential equations (9.1) and (9.2) in terms of a single differential equation (9.15) for the shear modulus μ as a function of $v = K/\mu$. This equation reads:

$$\frac{d\mu}{\mu} = \frac{27v^2(4+\alpha\pi)+3v(64+9\alpha\pi)+(64+6\alpha\pi)}{135v\left[v^3+\left(\frac{162-27\alpha\pi}{135}\right)v^2-\left(\frac{72+27\alpha\pi}{135}\right)v-\left(\frac{64+6\alpha\pi}{135}\right)\right]}dv \tag{9.2.1}$$

The expression inside the square brackets in the denominator of this equation is a third-degree polynomial in v. For all possible values of the aspect ratio α, this polynomial has three different real roots, so equation (9.2.1) can be written formally as:

$$\frac{d\mu}{\mu} = \frac{27v^2(4+\alpha\pi)+3v(64+9\alpha\pi)+(64+6\alpha\pi)}{135v(v-\rho_1)(v-\rho_2)(v-\rho_3)}dv \tag{9.2.2}$$

where ρ_1, ρ_2 and ρ_3 are the three roots of the polynomial. In explicit form, the values of the roots are:

$$\rho_1 = 2r^{\frac{1}{3}}\cos\left(\frac{\theta}{3}\right)-\left(\frac{6-\alpha\pi}{15}\right),$$

$$\rho_2 = 2r^{\frac{1}{3}}\cos\left(\frac{\theta+2\pi}{3}\right)-\left(\frac{6-\alpha\pi}{15}\right)$$

and

$$\rho_3 = 2r^{\frac{1}{3}}\cos\left(\frac{\theta+4\pi}{3}\right)-\left(\frac{6-\alpha\pi}{15}\right),$$

where

$$r = \left[\frac{1}{27} \left(\frac{76 + 3\alpha\pi + \alpha^2\pi^2}{75} \right)^3 \right]^{\frac{1}{2}}$$

and

$$\theta = \arccos \left\{ \frac{1}{2} \frac{\left[448 + 216\alpha\pi + 9\alpha^2\pi^2 + 2\alpha^3\pi^3 \right]}{\left[76 + 3\alpha\pi + \alpha^2\pi^2 \right]^{\frac{3}{2}}} \right\}$$

The right-hand side of equation (9.2.2) can therefore be expanded as:

$$\frac{27v^2(4 + \alpha\pi) + 3v(64 + 9\alpha\pi) + (64 + 6\alpha\pi)}{135v(v - \rho_1)(v - \rho_2)(v - \rho_3)} = \frac{A_1}{v} + \frac{A_2}{(v - \rho_1)} + \frac{A_3}{(v - \rho_2)} + \frac{A_4}{(v - \rho_3)}$$

where the 4 constants A_1 to A_4 are given by:

$$A_1 = -\frac{(64 + 6\alpha\pi)}{135\rho_1\rho_2\rho_3}$$

$$A_2 = \frac{\frac{(4 + \alpha\pi)}{5}\rho_1^2 + \frac{(64 + 9\alpha\pi)}{45}\rho_1 + \frac{(64 + 6\alpha\pi)}{135}}{\rho_1(\rho_1 - \rho_2)(\rho_1 - \rho_3)}$$

$$A_3 = \frac{\frac{(4 + \alpha\pi)}{5}\rho_2^2 + \frac{(64 + 9\alpha\pi)}{45}\rho_2 + \frac{(64 + 6\alpha\pi)}{135}}{\rho_2(\rho_2 - \rho_1)(\rho_2 - \rho_3)}$$

and

$$A_4 = \frac{\frac{(4 + \alpha\pi)}{5}\rho_3^2 + \frac{(64 + 9\alpha\pi)}{45}\rho_3 + \frac{(64 + 6\alpha\pi)}{135}}{\rho_3(\rho_3 - \rho_1)(\rho_3 - \rho_2)}$$

Hence, equation (9.2.2) becomes:

$$\frac{d\mu}{\mu} = \frac{A_1 dv}{v} + \frac{A_2 dv}{(v - \rho_1)} + \frac{A_3 dv}{(v - \rho_2)} + \frac{A_4 dv}{(v - \rho_3)} \tag{9.2.3}$$

Equation (9.2.3) can be integrated to give:

$$\ln\left(\frac{\mu}{\mu_1}\right) = A_1 \ln\left(\frac{v}{v_1}\right) + A_2 \ln\left(\frac{v - \rho_1}{v_1 - \rho_1}\right) + A_3 \ln\left(\frac{v - \rho_2}{v_1 - \rho_2}\right) + A_4 \ln\left(\frac{v - \rho_3}{v_1 - \rho_3}\right)$$

Before the penny cracks are progressively added to the pore space, as the DEM method requires, it is assumed that the rock has an initial matrix porosity ϕ_0 due to the presence of spherical inclusions. Let K_1 and μ_1 denote the bulk and shear moduli, respectively, of a rock with porosity ϕ_0. These moduli can in principle be calculated using the DEM method as well. Then if $v_1 = K_1 / \mu_1$ the last equation integrates to give:

$$\mu = Cv^{A1}\left(v - \rho_1\right)^{A2}\left(v - \rho_2\right)^{A3}\left(v - \rho_3\right)^{A4} \tag{9.2.4}$$

where

$$C = \frac{\mu_1}{v_1^{A1}\left(v_1 - \rho_1\right)^{A2}\left(v_1 - \rho_2\right)^{A3}\left(v_1 - \rho_3\right)^{A4}}$$

Now that the relationship between K and μ has been fixed, it is possible to solve any of the original differential equations for the porosity y as a function of K/μ, as follows:
Equation (9.1) with $K_f = 0$ and

$$P = \frac{K(3K + 4\mu)}{\alpha\pi\mu(3K + \mu)}$$

reads:

$$(1-y)\frac{dK}{dy} = -\frac{K}{\alpha\pi}\left(\frac{K}{\mu}\right)\frac{\left(3\dfrac{K}{\mu} + 4\right)}{\left(3\dfrac{K}{\mu} + 1\right)} \tag{9.2.5}$$

If $v = K/\mu$ as before, then:

$$\frac{dK}{dy} = \frac{dK}{dv}\frac{dv}{dy}$$

and equation (9.2.5) becomes

$$\frac{dy}{(1-y)} = -\frac{\alpha\pi}{v}\frac{\dfrac{dK}{dv}(3v+1)}{K(3v+4)}dv$$

or equivalently

$$\frac{dy}{1-y} = -\left(\frac{\alpha\pi}{9}\right)\frac{(3v+1)(3v+2)}{v(v-\rho_1)(v-\rho_2)(v-\rho_3)}dv \tag{9.2.6}$$

Here,

$$-\left(\frac{\alpha\pi}{9}\right)\frac{(3v+1)(3v+2)}{v(v-\rho_1)(v-\rho_2)(v-\rho_3)} = \frac{B_0}{v} + \frac{B_1}{v-\rho_1} + \frac{B_2}{v-\rho_2} + \frac{B_3}{v-\rho_3}$$

with

$$B_0 = \left(\frac{2\alpha\pi}{9}\right)\frac{1}{\rho_1\rho_2\rho_3}$$

$$B_1 = -\frac{\alpha\pi}{9}\frac{(3\rho_1+1)(3\rho_1+2)}{\rho_1(\rho_1-\rho_2)(\rho_1-\rho_3)}$$

$$B_2 = -\frac{\alpha\pi}{9}\frac{(3\rho_2+1)(3\rho_2+2)}{\rho_2(\rho_2-\rho_1)(\rho_2-\rho_3)}$$

and

$$B_3 = -\frac{\alpha\pi}{9}\frac{(3\rho_3+1)(3\rho_3+2)}{\rho_3(\rho_3-\rho_1)(\rho_3-\rho_2)}$$

It is a straightforward matter to integrate equation (9.2.6). The final result is:

$$-\ln(1-y) = B_0\ln v + B_1\ln(v-\rho_1) + B_2\ln(v-\rho_2) + B_3\ln(v-\rho_3) + C'$$

which is the same as equation (9.17) above.

APPENDIX 9.3: THE PROBABILITY THAT PENNY CRACKS WILL BE INTERCONNECTED

A fundamental assumption made in this chapter is that penny cracks can be treated as fractures and that they form an interconnected network. The existence of such a network would entail that the rock is permeable, even if the overall porosity is low.

Here, a semi-quantitative analysis will be performed in an attempt to justify the hypothesis that there is a relationship between the aspect ratio of the penny cracks and the probability that the network of fractures is interconnected. It is necessary to first make a number of simplifying assumptions, as listed below:

1. The problem will be treated in two dimensions only. Spherical pores will be represented by circles, and penny cracks – which are in reality ellipsoids – by ellipses.
2. However, because the cross-sections of these ellipses are very elongated (typically with aspects ratios less than 0.1), we will replace the ellipses with lines of infinitesimal thickness. Two penny cracks will therefore count as connected if there is an intersection between the two lines.
3. We will analyse the interconnectivity of just two spheres and just two lines. The ruling assumption is that if the interconnectivity of two lines is much greater than the interconnectivity of two spheres, the interconnectivity of n lines will be greater than the interconnectivity of n spheres.

4. Although it is reasonable to regard two lines as being interconnected if they intersect at any point, the interconnectivity of two spheres is more difficult to define. Figure 9.3.1 illustrates the definition of the interconnectivity of two spheres that will be adopted here. The first sphere is represented by the circle with black fill at the centre of the plot. Four possible positions of a second circle are shown. The two circles will be treated as interconnected if the centre of the second circle falls inside the red circle shown.
5. The two circles are assumed to have the same radius R and the two lines to have the same length L.

The planar region containing either the two circles or the two lines will have an area A_T. The porosity of the corresponding "rock", in the case where circles act as pores, is then $2\pi R^2/A_T$. On the other hand, if the lines act as the pores, the porosity is given by $2Ldx/A_T$, where dx is the thickness of the lines. Since the aspect ratio is $\alpha = dx/L$, the porosity in the second case can equivalently be written as $2L^2\alpha/A_T$. By hypothesis, the two systems have the same porosity, so $\pi R^2 = L^2\alpha$.

Figure 9.3.2 illustrates the circumstances under which the two lines would intersect. The two lines are represented by the vectors **AB** and **BC**, both with length L. The acute angle between the two vectors is denoted by THETA in the figure, and by θ below. If θ and the vector **BC** remain constant, the two vectors will intersect if the tail point A of the vector **AB** lies inside the parallelogram ABCD shown in the figure. So the probability that both lines will intersect is given by the area of the parallelogram divided by the total area and will be:

$$P_{line} = \frac{L^2 \sin(\theta)}{A_T} \tag{9.3.1}$$

The angle θ is a random variable which may have any value between 0 and $\pi/2$. The average value of the $\sin(\theta)$ is given by:

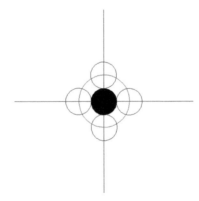

Figure 9.3.1 The black filled circle is the first circle. The unfilled black circles represent four possible positions of the second circle. If the centre of the second circle falls inside the red circle, the first and second circles are said to be interconnected.

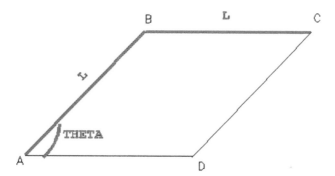

Figure 9.3.2 If vector BC and the angle THETA are kept fixed, the vectors AB and BC will intersect if the tail A of AB lies anywhere inside the parallelogram ABCD.

$$av[\sin(\theta)] = \frac{2}{\pi} \int_0^{\pi/2} \sin(\theta)d\theta = \frac{2}{\pi}$$

The average value of P_{line} is therefore given by:

$$avP_{line} = \frac{2L^2}{\pi A_T}. \qquad (9.3.2)$$

To simplify the notation we will continue to denote this average as P_{line}, although it should be remembered that it now represents an average probability.

Incidentally, this is a version of an old and famous problem known as "Buffon's needle". In its original formulation, Buffon's needle problem considers a series of parallel lines spaced a distance d apart and asks for the probability that a needle of length $L < d$ dropped at random would intersect one of the lines. The answer to this problem is:

$$P = \frac{2L}{\pi d}$$

Buffon's problem can be converted into the problem we are considering by imagining that each of Buffon's lines is subdivided into needles of length L and that the spacing d is chosen so that the area per needle is A_T. This can be done by requiring the rectangle of length L and width d about each needle to have area A_T. Then $d = A_T/L$ and the required probability is:

$$P = \frac{2L^2}{\pi A_T},$$

which is identical to equation (9.3.2).

In the case of circular pores, the probability of a connection between the pores is given by:

$$P_{circ} = \frac{4\pi R^2}{A_T} \tag{9.3.3}$$

because the red circle in Figure 9.3.1 has radius $2R$ and therefore area $4\pi R^2$.

After dividing equations (9.3.2) by (9.3.3) and bearing in mind that $\pi R^2 = L^2\alpha$, the final result is:

$$\frac{P_{line}}{P_{circle}} = \frac{1}{2\pi\alpha} \tag{9.3.4}$$

From equation (9.3.4), we see that the interconnectivity between two lines is in general much larger than the interconnectivity between two circles, as is illustrated by the sample ratios in the following table:

α	P_{line}/P_{circ}
0.1	1.59
0.01	15.9
0.002	79.6
0.001	159.1

The argument that has been outlined here can be summarized as follows:

1. When using the DEM model, new inclusions are added at random. Each added inclusion may replace the original solid material or previous inclusions, either partially or totally. Under these conditions, the smaller the aspect ratio of the inclusions, the greater will be the probability of an intersection between any two inclusions (as has just been shown).
2. If it is accepted that the principle stated in point 1) – namely, that the probability of intersection of two inclusions increases with decreasing aspect ratio – extends equally to the case of n inclusion, then for smaller aspect ratios there will be a greater probability of a network of interconnected fractures.
3. We claim that it is reasonable to suppose that the permeability of a fractured rock is related to the probability that the fractures will be interconnected, provided that all other factors remain the same, particularly the length of the fractures. This is a heuristic postulate which we have not demonstrated directly, but it underpins the importance of much of the analysis presented in this final chapter.

9781032134956